RESUMES
FOR
SCIENTIFIC
AND
TECHNICAL
CAREERS

Professional Resumes Series

RESUMES FOR SCIENTIFIC AND TECHNICAL CAREERS

The Editors of
VGM Career Horizons

Second Edition

VGM Career Horizons
NTC/Contemporary Publishing Group

Library of Congress Cataloging-in-Publication Data

Resumes for scientific and technical careers / the editors of VGM Career
    Books ; revised by Kathy Siebel.—2nd ed.
       p.   cm. — (VGM professional resumes series)
    Rev. ed. of: Resumes for scientific and technical careers. c1993.
    ISBN 0-8442-2925-3
    1. Scientists—Employment.   2. Engineers—Employment.
  3. Résumés (Employment).   I. Siebel, Kathy.   II. VGM Career Books.
  III. Series.
  Q147.R47   1999
  650.14—dc21   99-35620
                                          CIP

We would like to acknowledge the assistance of Kathy Siebel
in compiling and editing this book.

Interior design by City Desktop Productions, Inc.

Published by VGM Career Books
A division of NTC/Contemporary Publishing Group, Inc.
4255 West Touhy Avenue, Lincolnwood (Chicago), Illinois 60712-1975 U.S.A.

International Standard Book Number: 0-8442-2925-3

00 01 02 03 04 05 VL 20 19 18 17 16 15 14 13 12 11 10 9 8 7 6 5 4 3 2

Contents

Introduction

Your resume is your first impression on a prospective employer. Though you may be articulate, intelligent, and charming in person, a poor resume may prevent you from ever having the opportunity to demonstrate your interpersonal skills, because a poor resume may prevent you from ever being called for an interview. While few people have ever been hired solely on the basis of their resume, a well-written, well-organized resume can go a long way toward helping you land an interview. Your resume's main purpose is to get you that interview. The rest is up to you and the employer. If you both feel that you are right for the job and the job is right for you, chances are you will be hired.

A resume must catch the reader's attention yet still be easy to read and to the point. Resume styles have changed over the years. Today, brief and focused resumes are preferred. No longer do employers have the patience, or the time, to review several pages of solid type. A resume should be only one page long, if possible. Time is a precious commodity in today's business world and the resume that is concise and straightforward will usually be the one that gets noticed.

Let's not make the mistake, though, of assuming that writing a brief resume means that you can take less care in preparing it. A successful resume takes time and thought, and if you are willing to make the effort, the rewards are well worth it. Think of your resume as a sales tool with the product being you. You want to sell yourself to a prospective employer. This book is designed to help you prepare a resume that will help you further your career—to land that next job, or first job, or to return to the workforce after years of absence. So, read on. Make the effort and reap the rewards that a strong resume can bring to your career. Let's get to it!

The Elements of a Good Resume

A winning resume is made of the elements that employers are most interested in seeing when reviewing a job applicant. These basic elements are the essential ingredients of a successful resume and become the actual sections of your resume. The following is a list of elements that may be used in a resume. Some are essential, some are optional. We will be discussing these in this chapter in order to give you a better understanding of each element's role in the makeup of your resume:

1. Heading

2. Objective

3. Work Experience

4. Education

5. Honors

6. Activities

7. Certificates and Licenses

8. Professional Memberships

9. Special Skills

10. Personal Information

11. References

The first step in preparing your resume is to gather information about yourself and your past accomplishments. Later you will refine this information, rewrite it in the most effective language, and organize it into the most attractive layout. First, let's take a look at each of these important elements individually.

Heading

The heading may seem to be a simple enough element in your resume, but be careful not to take it lightly. The heading should be placed at the top of your resume and should include your name, home address, and telephone numbers. If you can take calls at your current place of business, include your business number, since most employers will attempt to contact you during the business day. If this is not possible, or if you can afford it, purchase an answering machine that allows you to retrieve your messages while you are away from home. This way you can make sure you don't miss important phone calls. Always include your phone number on your resume. It is crucial that when prospective employers need to have immediate contact with you, they can.

Objective

When seeking a particular career path, it is important to list a job objective on your resume. This statement helps employers know the direction that you see yourself heading, so that they can determine whether your goals are in line with the position available. The objective is normally one sentence long and describes your employment goals clearly and concisely. See the sample resumes in this book for examples of objective statements.

The job objective will vary depending on the type of person you are, the field you are in, and the type of goals you have. It can be either specific or general, but it should always be to the point.

In some cases, this element is not necessary, but usually it is a good idea to include your objective. It gives your possible future employer an idea of where you are coming from and where you want to go.

The objective statement is better left out, however, if you are uncertain of the exact title of the job you seek. In such a case, the inclusion of an overly specific objective statement could result in your not being considered for a variety of acceptable positions; be sure to incorporate this information in your cover letter instead.

Work Experience

This element is arguably the most important of them all. It will provide the central focus of your resume, so it is necessary that this section be as complete as possible. Only by examining your work experience in depth can you get to the heart of your accomplishments and present them in a way that demonstrates the strength of your qualifications. Of course, someone just out of school will have less work experience than someone who has been working for a number of years, but the amount of information isn't the most important thing—rather, how it is presented, and how it highlights you as a person and as a worker will be what counts.

As you work on this section of your resume, be aware of the need for accuracy. You'll want to include all necessary information about each of your jobs, including job title, dates, employer, city, state, responsibilities, special projects, and accomplishments. Be sure to only list company accomplishments for which you were directly responsible. If you haven't participated in any special projects, that's all right—this area may not be relevant to certain jobs.

The most common way to list your work experience is in *reverse chronological order.* In other words, start with your most recent job and work your way backward. This way your prospective employer sees your current (and often most important) job before seeing your past jobs. Your most recent position, if the most important, should also be the one that includes the most information, as compared to your previous positions. If you are just out of school, show your summer employment and part-time work, though in this case your education will most likely be more important than your work experience.

The following worksheets will help you gather information about your past jobs.

WORK EXPERIENCE

Job One:

Job Title _____

Dates _____

Employer _____

City, State _____

Major Duties _____

_____

_____

_____

_____

_____

_____

_____

Special Projects _____

_____

_____

_____

Accomplishments _____

_____

_____

_____

_____

_____

_____

_____

Job Two:

Job Title _____

Dates _____

Employer _____

City, State _____

Major Duties _____

_____

_____

_____

_____

_____

_____

_____

Special Projects _____

_____

_____

_____

Accomplishments _____

_____

_____

_____

_____

_____

_____

_____

_____

Job Three:

Job Title _____

Dates _____

Employer _____

City, State _____

Major Duties _____

_____

_____

_____

_____

_____

_____

_____

_____

Special Projects _____

_____

_____

_____

Accomplishments _____

_____

_____

_____

_____

_____

_____

_____

_____

Job Four:

Job Title _____

Dates _____

Employer _____

City, State _____

Major Duties _____

_____

_____

_____

_____

_____

_____

_____

_____

Special Projects _____

_____

_____

_____

Accomplishments _____

_____

_____

_____

_____

_____

_____

_____

_____

Education

Education is the second most important element of a resume. Your educational background is often a deciding factor in an employer's decision to hire you. Be sure to stress your accomplishments in school with the same finesse that you stressed your accomplishments at work. If you are looking for your first job, your education will be your greatest asset, since your work experience will most likely be minimal. In this case, the education section becomes the most important. You will want to be sure to include any degrees or certificates you received, your major area of concentration, any honors, and any relevant activities. Again, be sure to list your most recent schooling first. If you have completed graduate-level work, begin with that and work in reverse chronological order through your undergraduate education. If you have completed an undergraduate degree, you may choose whether to list your high school experience or not. This should be done only if your high school grade point average was well above average.

The following worksheets will help you gather information for this section of your resume. Also included are supplemental worksheets for honors and for activities. Sometimes honors and activities are listed in a section separate from education, most often near the end of the resume.

EDUCATION

School One _____

Major or Area of Concentration _____

Degree _____

Dates _____

School Two _____

Major or Area of Concentration _____

Degree _____

Dates _____

Honors

Here you should list any awards, honors, or memberships in honorary societies that you have received. Usually these are of an academic nature, but they can also be for special achievement in sports, clubs, or other school activities. Always be sure to include the name of the organization honoring you and the date(s) received. Use the worksheet below to help gather your honors information.

HONORS

Honor One _____

Awarding Organization _____

Date(s) _____

Honor Two _____

Awarding Organization _____

Date(s) _____

Honor Three _____

Awarding Organization _____

Date(s) _____

Honor Four _____

Awarding Organization _____

Date(s) _____

Activities

You may have been active in different organizations or clubs during your years at school; often an employer will look at such involvement as evidence of initiative and dedication. Your ability to take an active role, and even a leadership role, in a group should be included on your resume. Use the worksheet provided to list your activities and accomplishments in this area. In general, you should exclude any organization whose name indicates the race, creed, sex, age, marital status, color, or nation of origin of its members.

ACTIVITIES

Organization/Activity _____

Accomplishments _____

_____

_____

Organization/Activity _____

Accomplishments _____

_____

_____

Organization/Activity _____

Accomplishments _____

_____

_____

Organization/Activity _____

Accomplishments _____

_____

_____

As your work experience increases through the years, your school activities and honors will play less of a role in your resume, and eventually you will most likely only list your degree and any major honors you received. This is due to the fact that, as time goes by, your job performance becomes the most important element in your resume. Through time, your resume should change to reflect this.

Certificates and Licenses

The next potential element of your resume is certificates and licenses. You should list these if the job you are seeking requires them and you, of course, have acquired them. If you have applied for a license, but have not yet received it, use the phrase "application pending."

License requirements vary by state. If you have moved or you are planning to move to another state, be sure to check with that state's board or licensing agency to be sure that you are aware of all licensing requirements.

Always be sure that all of the information you list is completely accurate. Locate copies of your licenses and certificates and check the exact date and name of the accrediting agency. Use the following worksheet to list your licenses and certificates.

CERTIFICATES AND LICENSES

Name of License _____

Licensing Agency _____

Date Issued _____

Name of License _____

Licensing Agency _____

Date Issued _____

Name of License _____

Licensing Agency _____

Date Issued _____

Professional Memberships

Another potential element in your resume is a section listing professional memberships. Use this section to list involvement in professional associations, unions, and similar organizations. It is to your advantage to list any professional memberships that pertain to the job you are seeking. Be sure to include the dates of your involvement and whether you took part in any special activities or held any offices within the organization. Use the following worksheet to gather your information.

PROFESSIONAL MEMBERSHIPS

Name of Organization _____

Offices Held _____

Activities _____

Dates _____

Name of Organization _____

Offices Held _____

Activities _____

Dates _____

Name of Organization _____

Offices Held _____

Activities _____

Dates _____

Name of Organization _____

Offices Held _____

Activities _____

Dates _____

Special Skills

This section of your resume is set aside for mentioning any special abilities you have that could relate to the job you are seeking. This is the part of your resume where you have the opportunity to demonstrate certain talents and experiences that are not necessarily a part of your educational or work experience. Common examples include fluency in a foreign language, or knowledge of a particular computer application.

Special skills can encompass a wide range of your talents—remember to be sure that whatever skills you list relate to the type of work you are looking for.

Personal Information

Some people include "Personal" information on their resumes. This is not generally recommended, but you might wish to include it if you think that something in your personal life, such as a hobby or talent, has some bearing on the position you are seeking. This type of information is often referred to at the beginning of an interview, when it is used as an "ice breaker." Of course, personal information regarding age, marital status, race, religion, or sexual preference should never appear on any resume.

References

References are not usually listed on the resume, but a prospective employer needs to know that you have references who may be contacted if necessary. All that is necessary to include in your resume regarding references is a sentence at the bottom stating, "References are available upon request." If a prospective employer requests a list of references, be sure to have one ready. Also, check with whomever you list to see if it is all right for you to use them as a reference. Forewarn them that they may receive a call regarding a reference for you. This way they can be prepared to give you the best reference possible.

Writing Your Resume

Now that you have gathered all of the information for each of the sections of your resume, it's time to write out each section in a way that will get the attention of whoever is reviewing it. The type of language you use in your resume will affect its success. You want to take the information you have gathered and translate it into a language that will cause a potential employer to sit up and take notice.

Resume writing is not like expository writing or creative writing. It embodies a functional, direct writing style and focuses on the use of action words. By using action words in your writing, you more effectively stress past accomplishments. Action words help demonstrate your initiative and highlight your talents. Always use verbs that show strength and reflect the qualities of a "doer." By using action words, you characterize yourself as a person who takes action, and this will impress potential employers.

The following is a list of verbs commonly used in resume writing. Use this list to choose the action words that can help your resume become a strong one:

administered	billed
advised	built
analyzed	carried out
arranged	channeled
assembled	collected
assumed responsibility	communicated

compiled	maintained
completed	managed
conducted	met with
contacted	motivated
contracted	negotiated
coordinated	operated
counseled	orchestrated
created	ordered
cut	organized
designed	oversaw
determined	performed
developed	planned
directed	prepared
dispatched	presented
distributed	produced
documented	programmed
edited	published
established	purchased
expanded	recommended
functioned as	recorded
gathered	reduced
handled	referred
hired	represented
implemented	researched
improved	reviewed
inspected	saved
interviewed	screened
introduced	served as
invented	served on

sold	tested
suggested	trained
supervised	typed
taught	wrote

Now take a look at the information you put down on the work experience worksheets. Take that information and rewrite it in paragraph form, using verbs to highlight your actions and accomplishments. Let's look at an example, remembering that what matters here is the writing style, and not the particular job responsibilities given in our sample.

WORK EXPERIENCE
Regional Sales Manager

Manager of sales representatives from seven states. Responsible for twelve food chain accounts in the East. In charge of directing the sales force in planned selling toward specific goals. Supervisor and trainer of new sales representatives. Consulting for customers in the areas of inventory management and quality control.

Special Projects: Coordinator and sponsor of annual food industry sales seminar.

Accomplishments: Monthly regional volume went up 25 percent during my tenure while, at the same time, a proper sales/cost ratio was maintained. Customer/company relations improved significantly.

Below is the rewritten version of this information, using action words. Notice how much stronger it sounds.

WORK EXPERIENCE
Regional Sales Manager

Managed sales representatives from seven states. Handled twelve food chain accounts in the eastern United States. Directed the sales force in planned selling toward specific goals. Supervised and trained new sales representatives. Consulted for customers in the areas of inventory management and quality control. Coordinated and sponsored the annual Food Industry Seminar. Increased monthly regional volume 25 percent and helped to improve customer/company relations during my tenure.

Another way of constructing the work experience section is by using actual job descriptions. Job descriptions are rarely written using the proper resume language, but they do include all the information necessary to create this section of your resume. Take the description of one of the jobs you are including on your resume (if you have access to it), and turn it into an action-oriented paragraph. Below is an example of a job description followed by a version of the same description written using action words. Again, pay attention to the style of writing, as the details of your own work experience will be unique.

WORK EXPERIENCE
Public Administrator I

Responsibilities: Coordinate and direct public services to meet the needs of the nation, state, or community. Analyze problems; work with special committees and public agencies; recommend solutions to governing bodies.

Aptitudes and Skills: Ability to relate to and communicate with people; solve complex problems through analysis; plan, organize, and implement policies and programs. Knowledge of political systems; financial management; personnel administration; program evaluation; organizational theory.

WORK EXPERIENCE
Public Administrator I

Wrote pamphlets and conducted discussion groups to inform citizens of legislative processes and consumer issues. Organized and supervised 25 interviewers. Trained interviewers in effective communication skills.

Now that you have learned how to word your resume, you are ready for the next step in your quest for a winning resume: assembly and layout.

Chapter 3

Assembly and Layout

A t this point, you've gathered all the necessary information for your resume, and you've rewritten it using the language necessary to impress potential employers. Your next step is to assemble these elements in a logical order and then to lay them out on the page neatly and attractively in order to achieve the desired effect: getting that interview.

Assembly

The order of the elements in a resume makes a difference in its overall effect. Obviously, you would not want to put your name and address in the middle of the resume or your special skills section at the top. You want to put the elements in an order that stresses your most important achievements, not the less pertinent information. For example, if you recently graduated from school and have no full-time work experience, you will want to list your education before you list any part-time jobs you may have held during school. On the other hand, if you have been gainfully employed for several years and currently hold an important position in your company, you will want to list your work experience ahead of your education, which has become less pertinent with time.

There are some elements that are always included in your resume and some that are optional. Following is a list of essential and optional elements:

Essential	Optional
Name	Job Objective
Address	Honors
Phone Number	Special Skills
Work Experience	Professional Memberships
Education	Activities
References Phrase	Certificates and Licenses
	Personal Information

Your choice of optional sections depends on your own background and employment needs. Always use information that will put you and your abilities in a favorable light. If your honors are impressive, then be sure to include them in your resume. If your activities in school demonstrate particular talents necessary for the job you are seeking, then allow space for a section on activities. Each resume is unique, just as each person is unique.

Types of Resumes

So far, our discussion about resumes has involved the most common type—the *reverse chronological* resume, in which your most recent job is listed first and so on. This is the type of resume usually preferred by human resources directors, and it is the one most frequently used. However, in some cases this style of presentation is not the most effective way to highlight your skills and accomplishments.

For someone reentering the workforce after many years or someone looking to change career fields, the *functional resume* may work best. This type of resume focuses more on achievement and less on the sequence of your work history. In the functional resume, your experience is presented by what you have accomplished and the skills you have developed in your past work.

A functional resume can be assembled from the same information you collected for your chronological resume. The main difference lies in how you organize this information. Essentially, the work experience section becomes two sections, with your job duties and accomplishments comprising one section and your employer's name, city, state, your position, and the dates employed making up another section. The first section is placed near the top of the resume, just below the job objective section, and can be called *Accomplishments* or *Achievements*. The second section, containing the bare essentials of your employment history, should come after the accomplishments section and can be titled *Work Experience* or *Employment History*. The other sections of your resume remain the same. The work experience section is the only one affected in the functional resume. By placing the section that focuses on your achievements first, you thereby draw attention to these achievements. This puts less emphasis on who you worked for and more emphasis on what you did and what you are capable of doing.

For someone changing careers, emphasis on skills and achievements is essential. The identities of previous employers, which may be unrelated to one's new job field, need to be downplayed. The functional resume accomplishes this task. For someone reentering the workforce after many years, a functional resume is the obvious choice. If you lack full-time work experience, you will need to draw attention away from this fact and instead focus on your skills and abilities gained possibly through volunteer activities or part-time work. Education may also play a more important role in this resume.

Which type of resume is right for you will depend on your own personal circumstances. It may be helpful to create a chronological and a functional resume and then compare the two to find out which is more suitable. The sample resumes found in this book include both chronological and functional resumes. Use these resumes as guides to help you decide on the content and appearance of your own resume.

Layout

Once you have decided which elements to include in your resume and you have arranged them in an order that makes sense and emphasizes your achievements and abilities, then it is time to work on the physical layout of your resume.

There is no single appropriate layout that applies to every resume, but there are a few basic rules to follow in putting your resume on paper:

1. Leave a comfortable margin on the sides, top, and bottom of the page (usually 1 to 1½ inches).

2. Use appropriate spacing between the sections (usually 2 to 3 line spaces are adequate).

3. Be consistent in the *type* of headings you use for the different sections of your resume. For example, if you capitalize the heading EMPLOYMENT HISTORY, don't use initial capitals and underlining for a heading of equal importance, such as Education.

CHRONOLOGICAL RESUME

Franklin Wu
5391 Southward Plaza
Walnut Creek, CA 94596
(510) 555-9008

JOB OBJECTIVE:

To obtain a position as a management optician in a fast-paced retail store.

EDUCATION:

Graduated Hayward Community College, Hayward, CA in June of 1991
Graduated North Central High School, Chicago, IL in June of 1989

WORK EXPERIENCE:

1995 - present

Great Spectacles, Walnut Creek, CA
Management Optician

Valley Vision, Pleasanton, CA
Management Optician and Frame Buyer

SPECIAL SKILLS:

People person; fashion styling experience; knowledge of adjustments, repairs, and
    fittings of glasses and contact lenses.

CERTIFICATION:

American Board of Optometry Certificate

SEMINARS:

Cal-Q Optics to prepare for licensing, 1997
Opti-Fair (annual, three-day seminars)

REFERENCES:

George Jones, O.D.
Great Spectacles, (510) 555-8941

Maria Lazar, Optician
Valley Vision, (510) 555-3726

FUNCTIONAL RESUME

PATRICIA WHITE
987 West 44th Street
Cheyenne, WY 82001
(307) 555-9872

PROFESSIONAL OBJECTIVE

Opportunity to demonstrate superior managerial ability and administrative decision-making skills in a nursing home environment.

SUMMARY OF QUALIFICATIONS

- High degree of motivation
- Ability and patience to train and develop office and professional staff
- Thorough knowledge of IBM PC, Lotus 1-2-3, WordPerfect, IBM 38
- 10-key by touch
- Dictation

EDUCATION

University of Wyoming, B.A. Business
Laramie, WY

EXPERIENCE

1994 to Present - Assistant Director, Longview Manor, Cheyenne, WY
1991 to 1994 - Business Manager, Mountain Top Nursing Home, Cheyenne, WY

REFERENCES

Excellent professional and personal references

4. Always try to fit your resume onto one page. If you are having trouble fitting all your information onto one page, perhaps you are trying to say too much. Edit out any repetitive or unnecessary information or shorten descriptions of earlier jobs. Be ruthless. Maybe you've included too many optional sections.

Don't let the idea of having to tell every detail about your life get in the way of producing a resume that is simple and straightforward. The more compact your resume, the easier it will be to read and the better an impression it will make for you.

In some cases, the resume will not fit on a single page, even after extensive editing. In such cases, the resume should be printed on two pages so as not to compromise clarity or appearance. Each page of a two-page resume should be marked clearly with your name and the page number, e.g., "Judith Ramirez, page 1 of 2." The pages should then be stapled together.

Try experimenting with various layouts until you find one that looks good to you. Always show your final layout to other people and ask them what they like or dislike about it, and what impresses them most about your resume. Make sure that is what you want most to emphasize. If it isn't, you may want to consider making changes in your layout until the necessary information is emphasized. Use the sample resumes in this book to get some ideas for laying out your resume.

Putting Your Resume in Print

Your resume should be typed or printed on good quality $8\frac{1}{2}'' \times 11''$ bond paper. You want to make as good an impression as possible with your resume; therefore, quality paper is a necessity. If you have access to a word processor with a good printer, or know of someone who does, make use of it. Typewritten resumes should only be used when there are no other options available.

After you have produced a clean original, you will want to make duplicate copies of it. Usually a copy shop is your best bet for producing copies without smudges or streaks. Make sure you have the copy shop use quality bond paper for all copies of your resume. Ask for a sample copy before they run your entire order. After copies are made, check each copy for cleanliness and clarity.

Another more costly option is to have your resume typeset and printed by a printer. This will provide the most attractive resume of all.

If you anticipate needing a lot of copies of your resume, the cost of having it typeset may be justified.

Proofreading

After you have finished typing the master copy of your resume and before you go to have it copied or printed, you must thoroughly check it for typing and spelling errors. Have several people read it over just in case you may have missed an error. Misspelled words and typing mistakes will not make a good impression on a prospective employer, as they are a bad reflection on your writing ability and your attention to detail. With thorough and conscientious proofreading, these mistakes can be avoided.

The following are some rules of capitalization and punctuation that may come in handy when proofreading your resume:

RULES OF CAPITALIZATION

- Capitalize proper nouns, such as names of schools, colleges, and universities, names of companies, and brand names of products.
- Capitalize major words in the names and titles of books, tests, and articles that appear in the body of your resume.
- Capitalize words in major section headings of your resume.
- Do not capitalize words just because they seem important.
- When in doubt, consult a manual of style such as *Words Into Type* (Prentice-Hall), or *The Chicago Manual of Style* (The University of Chicago Press). Your local library can help you locate these and other reference books.

RULES OF PUNCTUATION

- Use a comma to separate words in a series.
- Use a semicolon to separate series of words that already include commas within the series.
- Use a semicolon to separate independent clauses that are not joined by a conjunction.

- Use a period to end a sentence.

- Use a colon to show that examples or details follow that will expand or amplify the preceding phrase.

- Avoid the use of dashes.

- Avoid the use of brackets.

- If you use any punctuation in an unusual way in your resume, be consistent in its use.

- Whenever you are uncertain, consult a style manual.

Chapter 4

The Cover Letter

Once your resume has been assembled, laid out, and printed to your satisfaction, the next step before distribution is to write your cover letter. Though there may be instances where you deliver your resume in person, usually you send it through the mail. Resumes sent through the mail always need an accompanying letter that briefly introduces you and your resume. The purpose of the cover letter is to get a potential employer to read your resume, just as the purpose of your resume is to get that same potential employer to call you for an interview.

Like your resume, your cover letter should be clean, neat, and direct. A cover letter usually includes the following information:

1. Your name and address (unless it already appears on your personal letterhead).

2. The date.

3. The name and address of the person and company to whom you are sending your resume.

4. The salutation ("Dear Mr." or "Dear Ms." followed by the person's last name, or "To Whom it May Concern" if you are answering a blind ad).

5. An opening paragraph explaining why you are writing (in response to an ad, the result of a previous meeting, at the suggestion of someone you both know) and indicating that you are interested in whatever job is being offered.

6. One or two more paragraphs that tell why you want to work for the company and what qualifications and experience you can bring to that company.

7. A final paragraph that closes the letter and requests that you be contacted for an interview.

8. The closing ("Sincerely," or "Yours truly," followed by your signature with your name typed under it).

Your cover letter, including all of the information above, should be no more than one page in length. The language used should be polite, businesslike, and to the point. Do not attempt to tell your life story in the cover letter. A long and cluttered letter will only serve to put off the reader. Remember, you only need to mention a few of your accomplishments and skills in the cover letter. The rest of your information is in your resume. Each and every achievement should not be mentioned twice. If your cover letter is a success, your resume will be read and all pertinent information reviewed by your prospective employer.

Producing the Cover Letter

Cover letters should always be individualized, since they are always written to particular individuals and companies. Never use a form letter for your cover letter. Cover letters cannot be copied or reproduced like resumes. Each one should be as personal as possible. Of course, once you have written and rewritten your first cover letter to the point where you are satisfied with it, you can use similar wording in subsequent letters.

After you have typed your cover letter on quality bond paper, proofread it as thoroughly as you did your resume. Again, spelling errors are a sure sign of carelessness, and you don't want that to be a part of your first impression on a prospective employer. Make sure to handle the letter and resume carefully to avoid any smudges, and then mail both your cover letter and resume in an appropriately sized envelope. Be sure to keep an accurate record of all the resumes you send out and the results of each mailing, either in a separate notebook or on individual index cards.

Numerous sample cover letters appear at the end of this book. Use them as models for your own cover letter or to get an idea of how cover letters are put together. Remember, every one is unique and depends on the particular circumstances of the individual writing it and the job for which he or she is applying.

Now the job of writing your resume and cover letter is complete. About a week after mailing resumes and cover letters to potential employers, you will want to contact them by telephone. Confirm that your resume arrived, and ask whether an interview might be possible. Getting your foot in the door during this call is half the battle of a job search, and a strong resume and cover letter will help you immeasurably.

Chapter 5

Sample Resumes

This chapter contains dozens of sample resumes for people pursuing a wide variety of jobs and careers.

There are many different styles of resumes in terms of graphic layout and presentation of information. These samples also represent people with varying amounts of education and experience. Model your own resume after these samples. Choose one resume, or borrow elements from several different resumes to help you construct your own.

Jason Alexander
345 East 82nd Street
New York, NY 10028

Office: (212) 555-3654   **Home: (212) 555-9065**

OBJECTIVE   Senior MIS management position in an international firm with long-range personal growth potential.

SUMMARY   Fifteen years' MIS experience in developing large-scale commercial systems. Solid technical background in multi-language programming and system design with extensive user interfacing. Last ten years held increasingly important MIS management positions.

EDUCATION   Cornell University, B.S., Mathematics (1974), M.A., Statistics (1978).

TECHNICAL   Equipment: IBM, DOS/VSE, OS/MVS, CICS, TSO/ROSCOE, SPERRY, DMS/TIP.

Languages: BADIC, COBOL, FORTRAN, INQUIRE, MARK IV, RPG, SMP.

EXPERIENCE

1993 - Present   **DELOITTE & TOUCHE, NEW YORK, NEW YORK**
Assistant MIS Director, Systems & Programming (1995 - Present)

Direct a systems and programming organization of 40 MIS professionals in the development and enhancement of the firm's internal business system. Oversee major developments in the areas of client management, general ledger, accounts receivable, personnel, and partnership accounting.

Manager, MIS Applications Development (1993 - 1995)

Staffed and directed a Systems & Programming group of 20 MIS professionals in the design and development of a firm-wide on-line database to maintain the firm's client base and to track the professional consulting staff's time and expenses.

Page 1 of 3

Jason Alexander
Page 2 of 3

EXPERIENCE (cont.)

- Directed project definition and functional analysis phase of project life cycle.

- Recruited 10 full-time analysts and programmers to develop detailed systems design and specifications using top-down structured methodology.

- Coordinated the design and development of a complex database structure to support the on-line informational needs of the firm.

- Initiated the development of naming standards, program skeletons, reusable code, and macro routines to assist and standardize the program construction phase of development.

- Developed and instituted a basecase testing methodology for the comprehensive testing and quality assurance of the developed system.

1983 - 1993   **IBM CORPORATION, WHITE PLAINS, NEW YORK**
Manager, MIS Development Projects (1991 - 1993)

Coordinated migration and implementation of financial and administrative systems being developed by European operations for use in the Latin American affiliates.

- Defined and implemented migration strategy and procedure for all MIS activities, including software, testing, and affiliate training.

- Conceived, designed, and negotiated first-time maintenance and emergency procedures for ongoing production support between European development center and Latin American affiliates.

- Initiated, developed, and implemented on-line strategy on optimum cost-effective general guidelines and procedures for all CICS applications in Latin America.

EXPERIENCE (cont.)

Manager, Systems Projects (1986 - 1991)

Developed and managed increasingly complex systems culminating in the direction of two-year retail finance and leasing system.

- Directed and implemented front-end CICS system that reduced input from ten to three days.

- Expanded and revamped existing system to mechanize all edits/validations and provide timely management reporting.

Directed systems and programming staff of eight in development/maintenance of systems in areas of supply/service billing, equipment control, collection services, insurance loss, and vehicle asset tracking.

- Conceived, directed, and implemented modularized table system.

- Instituted "Block" system releases resulting in greater throughput of user requests and reduced direct overhead costs.

Senior Programming Consultant (1983 - 1986)

Project leader of five systems and programming personnel in maintenance and enhancement of monthly equipment billing system.

Organized MARK IV Coordinator function to support end users of marketing, finance, and service/ distribution in developing their own inquiries for management information. Designed and programmed MARK IV applications.

PABLO SANCHEZ

500 Briar Patch Road
Frankfort, KY 40601
(502) 555-8750

OBJECTIVE

Foreman or heavy equipment operator.

PROFESSIONAL EXPERIENCE

From June 1995 to Present

Scraper Superintendent, Kentucky Department of Transportation.

- Involved in extensive training program initiated by state to give hands-on training to operators and mechanical staff on Cat models 631E, D10N, and 16G.

- Trained operators in the loading sequence by utilizing the Chain Loading method.

- Supervised and directed scraper fleet on a daily basis.

- Supervised and operated dozer on earth fill dam project.

- Operated truck fleet on overburden removal in a gold mine.

From July 1990 to January 1995

Owner and Operator of trucking/construction company, Sanchez Construction.

- Worked on small- to medium-sized construction projects for private sector as well as Soil Conservation Service and U.S. Army Corps of Engineers.

- Assistant Quality Control official for Soil Conservation Service and U.S. Army Corps of Engineers on flood cleanup.

- Instituted my own maintenance program and also did most of the mechanical work.

- Trucking for such contractors as Mason Corp., Daniel Industries, SJ Almaden, PB Snyder, Kentucky Excavating, Anton Construction, Hugh Bowman Contracting, and Landfill Crossroads, Inc.

Page 1 of 2

Pablo Sanchez
Page 2 of 2

EQUIPMENT SKILLS
From 1986 to 1990

Caterpillar:	Dozers, Loaders, Scrapers, Excavators
Komatsu:	Dozers, Excavators
Clark:	Excavators
John Deere:	Excavators
Holland:	Belt Loaders

EDUCATION

Mason County Joint Vocational School

Diesel Mechanics and Welding

Graduated 1985

MEMBERSHIPS

Fraternal Order of Elks

Fraternal Order of Police, Lodge #55

International Union of Operating Engineers, Local 555,
    Frankfort, Kentucky

REFERENCES

Available upon request.

MARK F. FULTON

Current Address:
78 Prairie Road
Columbus, OH 43216
(614) 555-3981

Permanent Address:
26 Frenwood Road
Steubenville, OH 40605
(614) 555-1807

OBJECTIVE

To obtain a position in construction engineering and management.

EDUCATION

Ohio State University, Columbus, OH
Master of Science, Civil Engineering
Graduated May 1999

University of Dayton, Dayton, OH
Bachelor of Engineering, Civil Engineering
Graduated Cum Laude, May 1998

HONORS

Regents Fellowship, Ohio State University
Harry Long Memorial Prize, University
  of Dayton
Dean's List for six semesters, University
  of Dayton

EXPERIENCE

SUMMER 97

Ohio Department of Transportation,
Steubenville, OH
Engineering in Training 1
Worked with Resident Engineer's office.
Responsible for inspecting the construction of
a reinforced concrete bridge and roadway.
Made daily reports and kept detailed records
of contractor performance and progress.

SUMMER 96

Hayes, Jones, and Bowers, Steubenville, OH
Engineering Aide
Worked with professional engineers in the
design of roadways and structures.
Responsible for hydrologic aspects of
highway design. Participated on in-depth
bridge inspection.

Page 1 of 2

Mark F. Fulton
Page 2 of 2

EXPERIENCE (cont.)

SUMMER 95	**Hayes, Jones, and Bowers,** Steubenville, OH *Engineering Aide* Worked with professional engineers in design of wastewater treatment plants and sewer systems. Responsible for AutoCAD drawings and checking design calculations.
CERTIFICATION	Passed the April 1999 Engineer in Training exam.
REFERENCES	Prof. B. William Gebbs Ohio State University 80 Lawrence Hall Columbus, OH 43210 Prof. Roger K. Hadley University of Dayton P.O. Box 8020-C Dayton, OH 45469

PATRICIA BUTTERFIELD
1480 Dean Road
Sacramento, CA 95819
(916) 555-9306

OBJECTIVE: Desire a challenging and rewarding position in the environmental field.

EDUCATION: Bachelor of Science, University of California at San Diego, 1992. Aquatic Biology, Minor in Natural Resources.

EXPERIENCE: CALIFORNIA DEPARTMENT OF ENVIRONMENTAL MANAGEMENT, Sacramento, CA.

March 1999 to Present

Environmental Project Manager, State Cleanup Section, Environmental Response.

- Manage the cleanup of hazardous waste sites.

- Contract the disposal of hazardous materials.

- Negotiate cleanup issues with PRPs.

- Conduct field sampling and contractor overview.

July 1998 to March 1999

Environmental Manager, Facilities Planning Section, Water Management.

- Implemented outreach program for building of treatment facilities in small communities.

- Performed primary and secondary review of facilities planning documents for the construction grants program.

- Interfaced with engineering firms and evaluated project costs.

April 1994 to July 1998

Environmental Scientist, Permits Section, Water Management.

- Reviewed program documents and managed implementation of municipal programs.

- Aided in the implementation of the state pretreatment program.

- Wrote permits for industrial users, NPDES, and land application.

- Conducted audits of municipal pretreatment programs and inspected manufacturing facilities.

Page 1 of 2

Patricia Butterfield
Page 2 of 2

EXPERIENCE (cont.):

LAKE SHASTA RECREATION AREA, Shasta, CA.

April 1993 to
September 1993   **Park Naturalist.**

- Managed the activities of park nature center.

- Conducted environmental programs for park visitors.

AFFILIATIONS: Institutes of Hazardous Materials Management
National Wildlife Federation
Water Pollution Control Federation

REFERENCES: Available upon request

Guy Fortune

Present Address
450 W. Pontiac Lane
East Lansing, MI 48824
(517) 555-6534

Permanent Address
1929 Ford Drive
Detroit, MI 48183
(313) 555-4631

Objective   Obtain engineering employment involving design and construction of roads, bridges, and associated structures.

Experience   Senior Class Project: Design of Concrete Slab Road Test Facility. Spring 1999

Appleby, Inc., Detroit, MI
Skilled Laborer/Supervisor
Site surveys and layouts, excavation, concrete placement and finishing, plumbing, carpentry, and equipment operation.
Summers 1996 - 1998

Lowell Construction, Detroit, MI
General Laborer
Concrete placement and finishing, carpentry, and excavation.
Summer 1993

Parr Marketing Co., Detroit, MI
Sales Representative. 12/96 - 1/97

L.D. Jones & Sons, East Lansing, MI
Retail Lumber Yard Helper. 6/98 - 12/98

Prairie Cycles, East Lansing, MI
Bicycle Repair Technician. 10/97 - 12/97

Education   Michigan State University, 5/99
B.S. in Civil Engineering
Dean's List, 1996 - 1997

University of Michigan, 6/98
Major - Electrical Engineering
Dean's List, 1994
Certified Engineer in Training, June 19, 1999

JOHN K. LAI
_____

20 West Concord Street
Dover, NH 03820
(603) 555-1703

EDUCATION: B.S., Civil Engineering, University of New Hampshire

CAREER SUMMARY:

Extensive experience in program management on complex construction projects. Managed all phases of project administration, contract development, and claims negotiation. Proven knowledge and skills to interact with professionals, contractors, and labor personnel.

EXPERIENCE:

1997 - Present	Johnson, Inc.
Position:	Resident Engineer, North Shore Interceptor, Phase IV.
Location:	Concord, New Hampshire
Duties:	Supervise performance of construction contractors. Project includes tunnels, deep shafts, chambers, odor control structures, and appurtenant facilities.
1996 - 1997	Bechtel
Position:	Project Manager
Location:	Hanford, Washington
Duties:	Special consultant to the Department of Energy and Rockwell International for design and construction of underground and shafts facilities for storage of nuclear waste.
1994 - 1996	Bechtel
Position:	Project Manager Construction Services
Location:	Los Angeles, California
Duties:	Prepared division budgets, long-range plans, and project proposals. Assigned construction personnel to projects. Acted as area construction manager on the proposal for Los Angeles Subway Construction. Supervised preparation of procedures for construction of power generation plant and coal mine in China.

Page 1 of 2

John K. Lai
Page 2 of 2

REGISTRATION: Registered to work in New Hampshire

AFFILIATIONS: American Management Association
Society of American Military Engineers
Society of Mining Engineers

REFERENCES: Available upon request

NAME: Michael S. Flowers

ADDRESS: 4459 Palm Drive
                 Las Vegas, NV 89154

PHONE: (702) 555-8666

EDUCATION:

1965 - 1966   University of Nevada, Reno, NV
                 Business Administration Major

1967 - 1968   Truckee Meadows Community College,
                 Truckee, NV, Math Major

1968 - 1973   International Business School, Civil
                 Engineering Certificate

1991 - 1993   University of Southern California, Los Angeles,
                 CA, Master of Business Administration

EXPERIENCE:

DESERT CONSTRUCTION, INC., Las Vegas, NV

9/98 - Present   **Manager of Estimating and Engineering**
                 Oversee and manage all estimating and
                 engineering.

L.A. SMITH CONSTRUCTION CO., San Diego, CA

4/95 - 9/98   **Manager of Estimating and Engineering**
                 Heavy/Marine Division. Supervise division
                 engineer and estimating manager for heavy
                 and marine estimating and construction.
                 Market segments are heavy engineering,
                 northeast transit, and hazardous waste work.

                                              Page 1 of 2

Page 2 of 2
Michael S. Flowers

EXPERIENCE (cont.):

BROWNELL CORPORATION, Long Beach, CA

11/89 - 3/95   **Vice-President/Manager of Heavy Estimating**
Established joint venture procedures. Arranged complete estimates for joint ventures. Supervised estimators in estimating and bidding projects such as treatment plants, power plants, airport terminals, piers and docks, pumping plants, subways, and mass transit. Upgraded computer format for heavy estimating.

FLOWERS CONSULTING, Los Angeles, CA

5/86 - 11/87   **President**
Formed company to provide consulting services. Advised Southern California Water District in reconstructing a CPM schedule of a water treatment plant.

KAREN S. ADAMS
1685 MOUNTAIN DRIVE
TUCSON, AZ 85720
(602) 555-8960

PROJECT MANAGEMENT

Project management for medium-sized company. Working toward international construction management with an eye on environmental compatibility.

PROFESSIONAL EXPERIENCE

1998 - Present University of Arizona Physical Plant, Tucson, AZ
Position: Project Coordinator
Responsibilities include:

- Start-to-finish management of construction projects.

- Estimating, assembling technical teams, surveying and layout, some design/drafting/ACAD, specs, job supervision, and inspection.

- Redesign of local "problem" intersection.

Prior to 1998 Thirteen years in the construction industry, starting as a laborer in 1982, ending as a journeyman, formsetter, and concrete finisher. Experience as crew supervisor, tractor operator, job supervisor, and estimator. Other experience includes many other construction jobs, motorcycle mechanic experience, waitressing, and owning and operating a woodworking shop.

STRENGTHS AND CAPABILITIES

- Detail and big picture-oriented.

- Anticipating and solving problems.

- Very good with numbers in the field and with people.

- Bringign projects in on time and within budget without sacrificing quality.

page 1 of 2

Karen S. Adams
page 2 of 2

ACADEMIC BACKGROUND

1990 - 1994 Senior in Civil Engineering at the University of Arizona, Tucson, AZ. G.P.A. of 3.85. Outstanding Junior and Outstanding Service Awards for 1990. Honors recipient every year from 1991 - 1994. Coursework stressing construction management, environmental and geotechnical engineering, IBM and MAC computer literacy, and French. Graduated May 1994.

REFERENCES Available upon request.

JOSH C. CURTAIN

655 Kelton Avenue, Detroit, MI 48125 (313) 555-7300

OBJECTIVE A position that will utilize my academic background and experience in outside industrial sales.

EDUCATIONAL BACKGROUND

1990 - 1994 University of Windsor, Windsor, Ontario. Bachelor of Commerce. Concentrating in marketing, finance, and accounting.

PROFESSIONAL EXPERIENCE April 1996 to present: Sales Engineer/Outside Sales Representative, Washington Electric, Detroit, MI Responsible for outside sales of special machinery and project coordinator in the Southeastern Michigan, Ohio, and Ontario areas. Launch and follow progress of machines from design to testing. Act as liaison between client/customer and plant. Assist in pricing and custom design of all machines. Develop and design brochures as selling tool. Continuously search for new and innovative techniques to reach the company's specific market. Deal predominantly with the auto, food, heating, and cooling industries.

SPECIAL SKILLS Working knowledge of WordPerfect and Lotus 1-2-3. Capable of making cold calls. Very familiar with Detroit automotive market and related industries.

REFERENCES Available upon request.

VAJID L. SINGH

P.O. Box 1296
Stanford, CA 94309
(415) 555-0701

EDUCATION *Master of Business Administration, June 1999*
Stanford University, Stanford, CA

Bachelor of Technology, Civil Engineering, July 1993
Institute of Technology, New Delhi, India

EMPLOYMENT

August 1993
to July 1997
Field Engineer
Nardini Co., Ltd. (subsidiary of Ferguson Construction, Inc., United Kingdom)
New Delhi, India

- Managed construction sites as independent profit centers consistently achieving target margins.

- Supervised and directed work of four supervisors and 25 skilled workers.

- Prepared cost estimates and quantity surveys for contract bids and analyzed project proposals.

- Collected, analyzed, and interpreted data pertaining to financial and production performance of site as well as wrote reports to facilitate control.

- Negotiated and liaisoned on a regular basis with labor unions and clients.

September 1997
to June 1999
Graduate Teaching Assistant
Stanford University
Stanford, CA

- Graded, tutored, and advised students enrolled in Operations Management class.

page 1 of 2

EMPLOYMENT (cont.)

September 1997
to June 1999

Microcomputer Laboratory Monitor
Stanford University
Stanford, CA

- Guided and assisted students and faculty in the productive use of analytical, graphics, and word-processing PC software.

- Served more than 100 users during peak laboratory usage.

COMPUTER BACKGROUND

Proficient in use of following PC programs, operating systems, and languages: RBase for DOS, Statgraphics, SPSS-PC+, Lotus 1-2-3, Quattro Pro, Excel, Harvard Graphics, MS-DOS, Netware, Pascal, BASIC, FORTRAN.

AWARDS & HONORS

Awarded $5,000 fellowship in 1997 by Education Trust to pursue study in the U.S.
Member - Beta Gamma Sigma (national honor society for business students)

REFERENCES

Supplied on request

KENNETH H. CROTHERS

80 Cavendish Drive   (608) 555-9783
Madison, WI 53714   (608) 555-6034

EDUCATION

Bachelor of Arts of Architecture, December 1998
University of Wisconsin, Madison, WI

EXPERIENCE

Landscape Architect
April 1998 - Present
NANCY'S NURSERIES & GARDEN CENTER, INC., Madison, WI

- Conduct cost estimates.
- Develop designs for residential/commercial landscapes.
- Lead final design presentation for client.

Partner/Designer
January 1998 - April 1998
SELF-EMPLOYED HOME RENOVATOR, Milwaukee, WI

- Developed ideas for home interior renovations.
- Implemented ideas/designs according to local building codes.

Assistant Layout Supervisor
May 1997 - September 1997
BRADFORD MALLS, Milwaukee, WI

- Assisted in the planning and layout operation for new and existing mall and restaurant parking areas.

Site Surveyor
May 1997 - August 1998
K&R INSTALLATIONS, Milwaukee, WI

- Installed recreational decks.

page 1 of 2

MILITARY

United States Marine Corps
September 1993 - April 1998
Anti-Armor Team Leader and Explosives Expert

RELATED SKILLS

- Operate CAD system.
- Read wide variety of design prints.

REFERENCES

Available upon request.

WILLIAM LYON COOPER
4142 Telegraph Avenue, #9
Berkeley, CA 94709
(408) 555-7756

Work History

1998 - Present   **Environmental Systems Group,** Oakland, CA
Project Manager. Prepare environmental assessments, facilities plans, and specialized reports. Collect data through research and field surveys. Conceptual design of wastewater treatment systems.

1994 - 1998   **Oregon State University Bookstore,** Corvallis, OR
Began as cashier, advanced to clerk-special order; further advancement to assistant branch manager. Ordered stock and inventory supplies; cleared and balanced daily sales; supervised hourly employees and customer relations. Worked 40 hours a week while attending college for three years.

Education

1993 - 1997   **Oregon State University,** Corvallis, OR
Bachelor of Science in Public Affairs, majors in Environmental Science and Environmental Affairs, requirements completed December 1997. Concentration grade point average of 2.9. Courses included Biology, Chemistry, Energy and the Environment, Environmental Techniques, Geology, Hydrogeology, Lake and Watershed Management, Law and Public Policy, Physics, and Urban Development.

References

Available on request.

LLOYD G. PRESCOTT

58 Mahwah Drive
Newark, New Jersey 07430
(201) 555-5297

EDUCATION

Purdue University, 1987

B.S. Industrial Management (minor - Industrial Engineering), GPA: 4.6/6.0

EMPLOYMENT SUMMARY

Approximately 15 years' experience

• Industrial Engineering
• Manufacturing and Process Engineering

EMPLOYMENT

8/98 to Present	**Central Engineering**, Newark, NJ	
	Senior Industrial Engineer	
1/98 to 8/98	**Newark Steel and Wire Company**, Newark, NJ	
	Contracted Industrial Engineer	
7/97 to 12/97	**Chicago Engineering**, Chicago, IL	
	Contracted Industrial Engineer	
4/94 to 5/97	**Northern Design**, Chicago, IL	
	Industrial Engineer	
9/87 to 4/94	**Industrial Designers**, South Bend, IN	
	Industrial Engineer	

ACCOMPLISHMENTS

• Developed process for assembly and fabrication departments.
• Conceptualized and drew layouts and tooling for assembly and fabrication.
• Managed Material Review Board.
• Established annual budget and monthly staffing for line operations.
• Coordinated cost reduction program.

REFERENCES AVAILABLE

NOLAN BRIAN KERR

930 North Leland Road
Flint, Michigan 48502
(313) 555-7623

CAREER OBJECTIVE

Position that requires technical knowledge in the areas of design, testing, and reliability of mechanical and electrical systems in order to produce a quality product.

EDUCATION

Purdue University, West Lafayette, Indiana
Attended: August 1987 - June 1989
Bachelor and Associate degrees in Mechanical Engineering Technology

Indiana State University, Terre Haute, Indiana
Attended: August 1985 - May 1987

Education included 20 hours of electronics and 10 hours of computer programming.

WORK EXPERIENCE

Michigan Lighting, Manufacturing Engineer.
M.L. is a joint American and Japanese automotive lighting company. Experience includes these areas: component designs, thermoset and thermoplastic molding, tooling, material evaluation, assembly line set-ups, adhesive development, robot feasibilities, and customer/supplier contacts. Familiar with foreign and domestic manufacturing concepts. Received Taguchi Design of Experiments training.

page 1 of 2

Nolan Brian Kerr
page 2 of 2

WORK EXPERIENCE *(cont.)*

Mercury Lamp Division of Ford Motor Company,
Project Engineer.
Five years' experience in working with automotive lighting systems. Performed the following functions: developed tests and implemented changes from test car and laboratory data; set up inventory systems; maintained budget, timing, and payroll records on computer; designed hardware parts for lamps; coordinated prototype parts; and designed layouts for the Forward Lighting facility. Supervised laboratory technicians, published testing manuals and reports, performed costs savings analysis on computer, and developed systems to monitor product performance in the field.

REFERENCES

Furnished upon request.

DENNIS P. WARDEN
152 Hogarth Avenue, Apt. #7D
Flushing, NY 11367
(718) 555-5401
warden@aol.com

BUSINESS EXPERIENCE

4/96 - Present   Industrial Engineer, Needham Cable
Flushing, NY

- Assist with world-class manufacturing and cell technologies implementation for rapid and continual improvement.

- Designed and programmed computerized suggestion system using dBase III, reducing clerical duties.

- Coordinated cost reduction program.

10/93 - 9/95   Plant Design Engineer, Coastal Cable
New Orleans, LA

- Justified and submitted cost reduction projects.

- Served as project engineer for OSHA safety project supervising completion of the project on time and under budget.

- Designed spreadsheet and database programs to create monthly master production schedules and material requirements for capacity analysis and JIT planning.

6/92 - 8/93   Plant Industrial Engineer, Lawrence Metals
Tampa, FL

- Established engineering standards to increase productivity.

- Reduced scrap and rejected shipments by writing detailed process sheets for operator use.

EDUCATION

Columbia University, 1991
School of Engineering and Applied Science
B.S. in Mechanical Engineering

Curtis S. Peeler
86 El Camino Real
Austin, Texas 75090

Office: (214) 555-7690   **Home: (214) 555-8929**

OBJECTIVE: MIS management position with upward potential.

SUMMARY: Outstanding track record:

- Management acceptance
- Vendor negotiations
- Systems/Programming/Operations
- Plan and policy formation

TECHNICAL: DL1/MS, DOS/VSE, IBM 37U/43xx, CICS, OS/MVS,
                     Cobol, Fortran, Mark IV, Datamanager.

APPLICATIONS: Distribution, finance, manufacturing, marketing,
                     operations.

EXPERIENCE:

1987 - Present   **IBM Corporation**
                     Austin, Texas
                     Multinational Manager of Business Systems

                     Managing multinational staff on enhancement and
                     implementation of equipment order, control and
                     invoicing COBOL systems developed by IBM in Europe
                     for Latin American operations.

- Evaluated and recommended acquisition
- Prepared cost/benefit ROI
- Created project organization requirements

page 1 of 2

Curtis S. Peeler
page 2 of 2

EXPERIENCE (cont.):

Manager of Business Systems (1989 - 1994)

Directed project managers on joint applications development for major subsidiaries. Retained Manager/Systems Applications responsibility.

- Programmed, purchased, or transferred applications via formal development and project methodologies
- Conducted overseas management reviews and sold improvement ideas
- Established centralized systems development staff in major local company

Manager of Systems Application (1987 - 1989)

Established and managed systems, programming, and operations group for HQ.

REFERENCES AVAILABLE

SHER POONA
48 Lawrence Hall
University of California
Berkeley, CA 94720
(510) 555-0568

OBJECTIVE   Development of computer and communication networks

EDUCATION   University of California at Berkeley, Berkeley, CA
M.S. in Electrical Engineering, December 1998
G.P.A.: 3.8/4.0

Chambal Regional College of Engineering, Kanpur, India
B.S. in Electrical Engineering, May 1997
Senior Project: Simulated a PC-based protection relay and
verified various algorithms for line and phase faults on power
transmission lines.

One-month industrial training at Delhi Electronics Limited,
Lahore, India, in the areas of data processing,
communication, and electronics.

BACKGROUND   Design, modeling, and analysis of centralized and distributed
networks, routing and flow algorithms, switching techniques,
multiple access for broadcast networks, data communication
hardware and software, packet-stitched and circuit-switched
networks, and satellite and local area networks.

Scientific Programming on VMS, UNIX, and MS-DOS in C,
Pascal, FORTRAN 77, Basic, and Assembly languages.

EXPERIENCE   **Adjunct Lecturer,** Department of Computer Science,
University of California at Berkeley. Instructor for an
undergraduate course in FORTRAN programming.
(1/98 to 7/98)

ACTIVITIES   **Treasurer,** IEEE Student Chapter, Kanpur, India
(1990 - 1991)
Coordinator, National Symposium on Applications of
Telecommunication in the Indian context, Kanpur, India
(September 1998)

JAMES B. WEITZMAN

980 CARPENTER ROAD
BUFFALO, NEW YORK 14208
(716) 555-3026

Summary
Desire estimator position with room for advancement. Salary requirements are open. References and photo portfolio are available.

Employment History

8/98 - Present
Baldwin Lumber Company, Buffalo, New York
Estimator/Designer

Prepare takeoffs of roof and floor trusses for single-family and multifamily dwellings as well as commercial and institutional structures. Work requires the ability to read and understand all types of architectural drawings and to develop working roof or floor designs that meet the architect's requirements and the owner's budget.

1/96 - 2/98
GMA Builders, Inc., Rochester, New York
Estimator/Drafter/Computer Operator

Estimated and designed custom homes. Developed an estimating method using the computer to increase efficiency and accuracy of residential estimating.

Education

12/95
Rochester Technical Institute, Rochester, New York
Associate Degree, with Honors
Architectural Engineering Technology

5/93
Batavia Consolidated High School, Rochester, New York
Graduated in top 20% of class
Coursework in drafting and woodworking

CHARLES W. WHITE
45 Cedar Pines Lane
Logan, Utah 84322
(801) 555-7516

OBJECTIVE   Capitalize on my experience in surveying and develop new skills in related fields.

EDUCATION   Utah State University, Logan, Utah
Bachelor of Science Degree in Earth Science,
May 1997

AREAS OF   Physical Geography   Meteorology
KNOWLEDGE   Chemistry   Geomorphology
Structural Geology   Glacial Geology
Mineralogy   Oceanography
Calculus   Physics
Wave Optics   Petrology
Paleontology   Astronomy

WORK EXPERIENCE

1/95 - Present   L. HARVEY WILL, Logan, Utah - Party Chief

Responsible for three-person crews. Work involves new subdivisions, construction layout, grade work, roads, boundary surveys, stakeouts, and title and deed research.

3/94 - 1/95   PRICE AND CASPER, Ogden, Utah - Party Chief
Performed surveys of residential and commercial property.

9/93 - 3/94   L. HARVEY WILL, Logan, Utah - Transit Man

1/93 - 9/93   ROBERT BECK AND ASSOCIATES, Provo, Utah - Assistant Surveyor

Participated in field work utilizing theodolite and transit, aerial photographs, tax maps, deeds, and sophisticated field instruments.

REFERENCES   Furnished upon request

PAUL J. RICHARDS
3300 Westwood Drive
Cuyahoga Falls, Ohio 44221
(216) 555-6929

OBJECTIVE

To attain a position as a designer/drafter with a highly aggressive architectural engineering firm.

SKILLS

Design and detailing of commercial mechanical, electrical, and plumbing systems. Architectural plans and details and site layout.

EXPERIENCE

4/98 to Present

THE INDUSTRIAL DESIGN GROUP, INC. Design development of mechanical, electrical, and plumbing systems within commercial projects. Produce final bid documents on multiple medias and AutoCad software. Develop construction details for architectural and engineering concepts. Responsible for pictorial sections used in site development and layout.

9/94 to 3/98

TACO BELL Supervised and evaluated the performance of twenty employees. Scheduled employees to maintain a productive operation. Responsible for payroll, daily and monthly accounting, and inventory control.

EDUCATION

Cleveland Institute of Applied Sciences Mechanical Design and Drafting Technology 12/98 Pursuing B.S. in Mechanical Engineering Technology

REFERENCES

Available upon request.

Scott Monroe
64 Fountain Lake Road
Gary, IN 46408
(219) 555-6823
monroe@netlink.com

Title: Electrical Engineer/Power Demand Maintenance

Education: B.S.E.E., 1991, Purdue University, G.P.A. of 3.5/4.0
M.S. Electrical Engineering, 1995, Northwestern University,
G.P.A. of 3.6/4.0

Experience:

BETHLEHEM STEEL, Bethlehem, PA

1994 - present AREA MANAGER-STEELMAKING RELIABILITY
Supervise salary and nonsalary positions. Developed and implemented a
new power demand system and new vacuum system.

1993 - 1994   STEEL OPERATIONS MAINTENANCE ENGINEER
Supervised combustion, electrical, and electronics engineering.
Investigated and solved electrical and mechanical problems.

1992 - 1993   SUPERVISOR-STEELMAKING RELIABILITY
Responsible for regular and preventative maintenance of steelmaking
equipment. Interfaced with support group and contractors.

1991 - 1992   ASSOCIATE MANUFACTURING ENGINEER
Investigated furnace transformer failures. Incorporated new technology.

ALLENTOWN STEEL, Pittsburgh, PA

1990 - 1991   ELECTRICAL ENGINEER
Installed quality control X-ray system. Designed and installed an automatic
conveyor system.

GOODYEAR TIRE & RUBBER COMPANY, Akron, OH

1988 - 1990   ELECTRICAL ENGINEER
Assisted in installation of new test wheel. Installed twelve new tire
presses.

References Available

EDGAR PETERS

9 De Soot Drive
Baton Rouge, LA 70805
(504) 555-1388

OBJECTIVE: Industrial engineering position with involvement in a manufacturing
               environment and opportunities to advance into production management.

EDUCATION: Master of Engineering
               Tulane University, 1996
               Major: Industrial Engineering

               Bachelor of Science
               Tulane University, 1995

EXPERIENCE:

1996 to present   Alexander Steel Company, Baton Rouge, LA

                  Associate Industrial Engineer
                  Provide identification and implementation of computer applications for
                  analysis and control. Activities include computer modeling and
                  economic and statistical analysis. Originated an operating change to
                  increase furnace hot-blast temperature. Developed diagnostic, routing,
                  quality control, and unit scheduling expert systems.

1995 to 1996   Packaging Systems, Inc., New Orleans, LA

                  Part-time Supervisor
                  Responsible for ten people in package-sort activities. Supervised,
                  evaluated, and trained sort and audit personnel.

1993 to 1995   Production Facilities, New Orleans, LA

                  Student Assistant
                  Assisted project managers in development of new production and test
                  facilities. Developed and documented procedures for initiating
                  component repairs.

REFERENCES: Available upon request.

BRYAN PULLMAN
43 BUFFALO BILL ROAD
OMAHA, NE 68129
(402) 555-5837

EDUCATION

- University of Nebraska at Lincoln, 1993, B.S. in geology.

- University of Nebraska at Lincoln, Graduate studies in geology, forestry, and natural resources.

PROFESSIONAL EXPERIENCE

Project Geologist
Buckman & Klein Engineering, Inc., 4/94 - present

- Manage environmental assessments for properties undergoing acquisition, divesture, or refinancing.

- Supervise and document underground storage tank removal and subsequent contaminated soil remediation.

- Designed soil venting systems, groundwater recovery/treatment systems, and bioremediation programs.

- Responsible for proposals, drill scheduling, material purchasing, invoicing, and client development on projects.

RELEVANT EXPERIENCE

Materials Engineering Technician
Maxwell Associates, 5/93 - 4/94

Engineering Assistant
Nebraska Department of Natural Resources Division of Water, 2/93 - 4/93

ACHIEVEMENTS

- Nebraska Academy of Science - Presented and published "The Formation of the Platte Sinkhole and the Drainage Effects on Platte and Stapleton Hollows."

- Geological Society of America - Presented research project on soil development at 1993 convention.

Lionel Dean
46 Bay Drive
Chicago, IL 60641
(312) 555-7862

CAREER
OBJECTIVE   A mechanical engineering position in a
               manufacturing/design evironment.

EDUCATION   B.S., Mechanical Engineering Technology
               University of Illinois at Urbana/Champaign
               Major G.P.A. - 4.8/5.0, December 1993

EXPERIENCE

1993 - Present UNITED STATES STEEL, East Chicago, IN

                HYDRAULIC/MAINTENANCE ENGINEER

                Hydraulic Engineer assigned as mechanical
                coordinator for revamp projects, and
                development of hydraulics schematics for
                revamp projects, and development of a
                hydraulic training program for mechanical
                maintenance personnel.

                Maintenance Engineer responsible for the
                design, procurement, and construction of
                various mechanical/structural projects.

                Duties include development and estimate of
                work, selection of an engineering contractor,
                scheduling and cost control, development of
                contractor bid packages, selection of a field
                contractor, and overall approval and
                supervision of field work.

                MAINTENANCE TURN SUPERVISOR

                Responsibilities included supervision of
                bargaining unit employees who maintained a
                flat-roll steel finishing facility.

Page 1 of 2

Lionel Dean
Page 2 of 2

EXPERIENCE (cont.)

1989 - 1993	UNIVERSITY GROUP, Urbana, IL
	MAINTENANCE SUPERVISOR
	Responsibilities included repair and operation of HVAC, building utilities, grounds, and their respective equipment.
AFFILIATIONS	Member of United Steelworkers Union #322 Member Building Trades Guild
REFERENCES	References available on request.

MARK CHANG
587 Flint Drive
Detroit, MI 48202
(313) 555-3578

Objective   To obtain a position as an engineer where I can apply my knowledge of digital circuit design, programmable controllers, and microprocessors.

Employment   Chrysler Data Systems
Detroit, MI
Systems Engineer - Initially worked with Electrical/HVAC group resolving computer problems, keeping inventory, and establishing the goals of the group. Now work as part of the Plant Systems group resolving problems, analyzing change requests, and writing troubleshooting documentation for an automated storage and retrieval system.
Dates: 8/97 to Present

Perkins Engineering
Flint, MI
Die Detailer - Responsibilities included drawing dimensional die details, making engineering changes to die drawings and details, and running blueprints.
Dates: 5/96 to 7/97

Education   Michigan State University, East Lansing, MI
Bachelor of Science degree in Electrical Engineering, May 1996

Passed the Professional Engineering Exam, April 1997

Affiliations   Institute of Electrical and Electronic Engineers

References   Available upon request.

WILLIAM J. SHAW

1810 Cedar Crest Boulevard
Evansville, IN 47722
(812) 555-1479

CAREER SUMMARY

More than twenty years' experience in manufacturing, production, and assembly of medium- and high-volume stamping and fabrication operations. Supervised activities in fabrication, stamping, welding, and finishing of automotive and agricultural equipment as well as appliances. Responsible for training, scheduling, safety, work quality, material movement, and discipline.

TECHNICAL QUALIFICATIONS

Experienced with JIT, MRP, Statistical Process control, and automated visual inventory/scheduling concepts. Proficient in high speed light stampings, transfer press operations, heavy stamped assemblies, welding, and testing instrumentation/procedures.

EMPLOYMENT HISTORY

Whirlpool Corporation, Evansville, IN   8/91 to present
Operations/Finishing Manager

Ford Motor Company, San Leandro, CA   6/79 to 8/91
Positions held: Supervisor of Cab Fabrication/Finishing, Maintenance Supervisor, and Finished Vehicle Assembly Supervisor.

EDUCATION

B.S. in Business Administration, Carnegie Mellon University, 1989

References available on request.

Rachel Schwartzman
125 College Way
Princeton, NJ 08545
schwartz@netlink.com

Objective   Mechanical engineering and/or programming position involving robotics or other electromechanical systems.

Education   **Princeton University**, Princeton, NJ
Major: Mechanical Engineering (GPA: 3.8/4.0)
1995 - Present
Won graduate research fellowship to perform independent robotics research.

Experience

Summer 1998   Programmer. **General Electric (Robotics Lab)**
Developed a robotic work cell using Linear-Motor-Robots to assemble washer-pumps at high production rates. Designed the gripper hardware for each task in the production cycle as well as all the fixturing for the parts being assembled.

Summer 1997   Mechanical Engineer. **Ford Motor Company**
Designed an integrated car set frame that used hydraulics to absorb as much of the impact energy in an accident. Also, designed a power window motor casing to facilitate robotic assembly of window systems on assembly line.

Summer 1996   Computer Programmer. **World Airways, Inc.**
Set up monthly customer mailing system/inventory database in D-Base and Pascal.

Summer 1995   Robotics/Computer Programmer. **Rutgers University**
Programmed Robotic Lab equipment and vision system for horticulture research.

Honors   General Electric Scholarship, full tuition
Member, Tau Beta Pi
Member, Phi Beta Kappa

ANA FONG
1260 PALMETTO DRIVE
ORLANDO, FL 32816
(305) 555-1318

OBJECTIVE

To utilize my management, marketing, and computer service experiences to make an immediate contribution as a member of a professional management team.

TECHNICAL EXPERTISE

Systems: CA/1, CA/7, CICS, MVS/XA, Vtam, SUPRA

Hardware: IBM 309X-308X, IBM 4331, StorageTek 4400 ACS

EMPLOYMENT SUMMARY

Corporate Operations Manager
Reynolds Corporation, Orlando, FL
1994 - Present

- Directed implementation of new data center and hired and trained operations and network personnel.
- Installed corporate telecommunications systems including PBXs and key systems.
- Coordinated hardware acquisitions and lease negotiations for all nationwide corporate facilities.
- Reduced printing costs by managing a project team through the analysis, design, development, and implementation of new printing systems and design procedures.

Operations Manager
Dade County Information Services Agency, Miami, FL
1991 - 1994

- Initiated automated problem resolution system resulting in reduction of problems and elimination of manual systems.
- Developed position titles and pay scales that resulted in identifiable career paths for operations personnel.

page 1 of 2

Ana Fong
page 2 of 2

EDUCATION

University of California at Los Angeles
B.S. in Computer Science, 1991

AFFILIATIONS

Florida Telecommunications User Association
Association for Computer Operations Managers
Reynolds Corporate Mentor for Partners in Education Program

REFERENCES

Available on request.

PETER DAWSON

420 Calumet Avenue, Gary, IN 46408   **219/555-6457**

OBJECTIVE

Position in process metallurgy/quality control.

CAREER SUMMARY

Fifteen years' service with a major manufacturer of flat-rolled and tubular products in various functional areas. Highly developed skills in work organization, metallurgical process control and applications, and expertise in finishing and management of basic manufacturing.

WORK EXPERIENCE

United States Steel Corporation, Gary, IN   1996 - present

Hot Mill Metallurgist (1998 - present)
Responsible for all aspects of hot strip mill quality including thermal practice, customer product and processing requirements, testing, and claims.

- Established new product/grade hot-rolling standards.
- Supervised hot strip mill quality control work force, including metallurgical turn supervisor, observers, and testing personnel.

High Carbon/Alloy Metallurgist (1996 - 1998)
Responsible for quality and thermal process control for all high carbon and alloy grades/products.

- Directed and coordinated slabbing and hot-rolling of customer conversion material.
- Established and developed standard operating and testing procedures for high-tech alloy application.

page 1 of 2

WORK EXPERIENCE (cont.)

United States Steel Corporation, Cleveland, OH 1988 - 1996

Shipping Supervisor (1994 - 1996)
Responsible for processing of various sizes, lengths, and grades of tubular products.

EDUCATION

B.S. Metallurgical Engineering, Purdue University,
West Lafayette, IN, 1988

AFFILIATIONS

American Steelworkers Association
Professional Metallurgical Engineers Guild
Steelworkers Union #458

REFERENCES

Available on request.

DAVID FORTSON

1038 Independence Drive • Dayton, OH 45469 • (513) 555-4429

PROFESSIONAL OBJECTIVE

Senior software engineer with responsibility for both small and large software projects from conceptual phase to finished product.

TECHNICAL EXPERTISE

Hardware: FORCE 68030, SUN Workstation, DEC VAX, HP 9000, IBM AT 386, DEC LS1 1140/44

Operating Systems: UNIX, VMS, VXWORKS, MS-DOS, VERSADOS

Languages: C, MACRO Assembler, FORTRAN, BASIC, ASYST

EXPERIENCE

Automatic Optical Systems
Columbus, Ohio
1998 - present

Senior Software Engineer - Responsible for detailed specifications, detailed designs, implementation, and maintenance of the software that controls and acquires data from a real-time experimental laser system. Key duties include providing software specification reviews for internal and external audiences.

- Integrated and implemented hardware/software system to acquire low- and high-speed data from different sensor channels.

- Implemented complex mathematical algorithms used to control the laser system and to provide diagnostic information. Worked under the direction of the chief research physicist.

Page 1 of 2

David Fortson
Page 2 of 2

EXPERIENCE (cont.)

Software, Inc.
Dayton, Ohio
1997

Senior Software Engineer - Performed software engineering services as a consultant for employers. Designed, developed, and debugged assembly language code for DEC VAX system performing on-line cellular telephone caller validation.

- Developed new code and modified existing code to decrease the time required to verify cellular telephone caller's identity.

OTX Y Systems
Dayton, OH
1987 - 1997

Senior Engineer - Responsible for design and implementation of high visibility national defense software projects.

- Worked as team leader and senior contributing member of several hardware/software design teams. Wrote software requirement specifications based upon DOD 2167A protocol. Comprehensive customer interface and design reviews were required.

- Designed, developed, and implemented software that was part of an HP9000-based Automatic Test System in conjunction with hardware designers and manufacturing engineering.

EDUCATION

B.S., Miami University, Oxford, OH

Currently enrolled in the MS Computer Science Software Engineering program, Ohio State University, Columbus, OH

REFERENCES

References are available and will be provided on request.

STEPHANIE MORRIS
11917 NORTH MERIDIAN
CARMEL, IN 46032
(317) 555-6463

OBJECTIVE:

Seeking a challenging position in the areas of systems analysis, database management, and programming that will utilize my technical and interpersonal skills.

EDUCATION:

University of Texas
Austin, Texas
B.S. in Computer Technology, Computer Information Systems
Minors in Business and Industrial Operations
Fall 1990, Fall 1995

WORK EXPERIENCE:

Carl James Associates, Indianapolis, Indiana
Associate (February 1997 to Present)
Converted all operations for an Indiana municipality from Burroughs ISAM/COBOL/RPG to HP3000 IMAGE/COOL. Redesigned, rewrote, developed, and implemented all applications and new development.

Mayflower Van Lines, Indianapolis, Indiana
Programmer/Analyst (February 1996 to February 1997)
Designed, developed, and implemented reporting applications for operations including financial and operational reporting. Responsible for PC hardware and software setup and support for 25 PCs. Provided user support for applications written in FOCUS on an Amdahl mainframe.

Hardware Wholesalers, South Bend, Indiana
Programmer/Analyst Trainee (summer 1995)
Programmed COBOL with IDMS, created an on-line application using CICS and COBOL, and wrote documentation and created new applications using Easytrieve Plus and Keymaster.

REFERENCES AVAILABLE

Michael G. Block   Home (617) 555-4813
75 Eldridge Court   Work (617) 555-6741
Cambridge, MA 02138

OBJECTIVE   To utilize my communication, problem-solving, and decision-making skills in
              a professional position that offers development and increasing levels of
              responsibility.

EDUCATION   Ivy Technical Institute - Associate Degree
              Material Requirements Planning Seminars - Certificate
              ITT Technical Institute - Certificate

EXPERIENCE

10/99 - Present   *Lincoln Engineering - Cambridge, MA*
                  Service Technician Assistant - Assist service technicians in installing heating
                  and air-conditioning units in various city-wide industrial and residential
                  applications. Provide pick-up and delivery service. Operate hydraulic forklift.
                  Use acetylene/oxygen cutting torch and other related trade tools.

4/94 - 9/99   *Taylor Components Group - Concord, NH*
                  Programmer Technician - Developed and maintained applications for various
                  departments. Created screen formats for program access using FOCUS
                  Report Writer language.

                  Trainer - Provided end users with working understanding of computer.
                  Taught in-house seminar on creating Bills-of-Material using Cullinet on-line
                  software package.

                  Help Service - Allocated, created, and deleted data sets for end users.
                  Provided troubleshooting assistance. Served as liaison between Management
                  Information Service and various departments.

                  Documentation - Prepared and provided end users with step-by-step
                  procedures for using computer. Prepared user manual for Bills-of-Materials
                  seminar.

4/89 - 4/94   *Taylor Components Group - Concord, NH*
                  Drafter - Prepared detailed drawings of parts from layouts and sketches
                  using standard drawing and drafting and measuring tools and instruments.

Robert L. Montgomery
46 Washington Boulevard
Albuquerque, New Mexico 87131
505-293-0510

EDUCATION: B.S. in Civil Engineering, 1985

EXPERIENCE: KAISER HEAVY CONSTRUCTION
                 4200 Santa Fe Drive
                 Albuquerque, New Mexico 87134

6/94 to present Senior Estimator in home office. Handle all
                 underground bids.

9/93 to 4/94   Special Projects Engineer on King Dam Project.
                 Supervised excavation of underground powerhouse.

                 MONTGOMERY, INC.
                 46 Washington Boulevard
                 Albuquerque, New Mexico 87131

1/91 to 9/93   Consultant to heavy construction contractors on
                 estimating, claim preparation, construction
                 management, and design of ground support systems.
                 Consultant to law firms on preparation of legal
                 proceedings involving construction claims.

4/88 to 1/91   Subcontractor doing pipe jacking, small tunnels,
                 bulkhead and stone revetments, structure grouting,
                 grading, and site-work.

                 R & G CONSTRUCTORS, INC.
                 8715 Highway 395
                 Reno, Nevada 89557

10/83 to 10/85   Project Engineer on Emigrant Gap Tunnel,
                 South Lake Tahoe, California.

2/85 to 4/88   Project Engineer on relief sewer in Reno.

Robert L. Montgomery
page 2 of 2

EXPERIENCE:
(cont.)

DILINGHAM-GRAVES-GRANITE
4424 N. 16th Street
Washington, D.C. 20223

10/83 to 2/85   General Engineering Consultant to Washington, D.C. Rapid Transit. Washington, D.C.

7/77 to 10/83   Started as Field Engineer and worked up to Resident Engineer. Involved in the following work:

- 2,400 feet of exploratory tunnels.

- 33,000 feet of tunnel driven by conventional methods in rock.

- 7,000 feet of tunnel driven using a boring machine.

- Spent 1 ½ years as designer on underground structures.

REGISTRATION: Professional Engineer in the following states:

New Mexico
District of Columbia
Nevada

**KEVIN G. ACKROYD
438 BEAVER DRIVE, APT. 67
UNIVERSITY PARK, PA 16802
(814) 555-9478**

EMPLOYMENT GOAL

Full-time employment in a medium-size company that does earthwork and/or heavy construction.

EDUCATION

Penn State University - Senior 1997 - 98, Graduate June 1998
Degree - Bachelor of Science in Construction Engineering Management

RELATED EXPERIENCE

ENGINEER INTERN - Williamsport Paving Co., Williamsport, PA, June to September 1997. Responsible for upkeep of the job-costing system on all projects and time and material billings. Some estimating, job supervision, signing, and laboring.

PROJECT OFFICER - 125th Engineer Battalion, PAANG, 1995 to present. The Officer-in-Charge of a road construction project and a haul-in project. Duties include coordination of materials, equipment, and direct supervision of project.

GRADE CHECKER/LABOR - White Construction Co., Wilkes-Barre, PA, July to September 1996. Gained experience in grade checking, pipe-laying, chip-seal, and flagging.

EQUIPMENT OPERATOR - Barron's Trenching, Altoona, PA, July to September 1995. Operated a CASE 580 Backhoe and 450 dozer in excavation for farm drainage systems and private contract work.

LEADERSHIP, ACTIVITIES, HONORS, & AWARDS

MEMBER - AGC Student Chapter, Penn State
COMMANDER - ROTC Drill Team
PLATOON LEADER - Heavy Equipment Platoon, National Guard
SCHOLARSHIP - AGC, 3-year, undergraduate

Leo Cervetto
18 Cliff Road
Portland, Oregon 97205
(503) 555-3546

PROFESSIONAL EXPERIENCE

COORDINATOR OF TECHNICAL SERVICES, ALLIANCE OREGON, INC.
December 1999 to present

In charge of asbestos program for more than 100 school buildings, involving review of existing asbestos management programs and extensive contact with school administrators in planning and implementation of timely and budget-sensitive management programs for environmental issues.

Regularly conducted field surveillance and inspection activities at all sites. Trained school personnel on various environmental issues including asbestos, lead, and radon.

Wrote company hazard communication, respiratory protection, and medical surveillance programs as well as standard operating procedures manuals for functions within the asbestos management program.

INDUSTRIAL HYGIENIST, ENVIRONMENTAL CONSULTANTS
March 1998 to October 1999

Project manager position involving coordination of industrial hygiene and asbestos-related projects: bulk sampling, technical report writing, abatement project specification development, environmental compliance monitoring, and project design. Supervised technician pool and client services.

Carried out wet chemistry procedures applicable to analysis of priority pollutants, both organic and inorganic, for solid and liquid matrices.

LABORATORY TECHNICIAN, BEAVER ANALYTICAL SERVICES
June 1997 to February 1998

Responsible for preparation of solid and liquid samples for the analysis of tetrachlorodibenzodioxin. Duties included sample check-in, solid/liquid extraction, various clean-up procedures, and standards preparation.

page 1 of 2

Leo Cervetto
page 2 of 2

EDUCATION

PORTLAND STATE UNIVERSITY, May 1997
    Bachelor of Science, Biology
    Chemistry minor

TRAINING/ACCREDITATION

OSHA compliance training

EPA-accredited building inspector, asbestos

EPA-accredited asbestos management planner, asbestos

Sampling and Evaluating Airborne Asbestos Dust certification

SPECIALTIES

All aspects of asbestos management in residential and commercial buildings

Air sample analysis by Polarized Light Microscopy

Indoor air quality evaluation

Industrial Hygiene sampling

WILBUR KENNEDY
619 Alameda Street
Santa Barbara, CA 93107
(805) 555-3428

SUMMARY

Eight years of nuclear operations experience in the U.S. Navy followed by four years of management experience and two years of design experience in the electronics industry.

EXPERIENCE

1993 - present

ACE ELECTRONICS GROUP, Santa Barbara, CA

Engineer   1996 - present

- Have designed more than ten optical sensors and industry controls.
- Have designed circuit board layouts.
- Supervise the production and testing of prototypes.
- Supervise the maintenance of engineering department records and drawings.
- Work with customers to design solutions to their applications.

Production Manager   1994 - 1996

- Made planning, controlling, and staffing decisions.
- Supervised production of sensors and oscillators.
- Selected and implemented software and hardware for computer accounting of inventory.

Technician   1993 - 1994

- Tested, adjusted, and repaired optical sensors and crystal oscillators.

page 1 of 2

Wilbur Kennedy
page 2 of 2

EXPERIENCE (cont.)

1985 - 1993   **U.S. NAVY**

Nuclear-powered Research Submarine   1989 - 1993

- Assigned as Interior Communications Officer and Computer Officer.
- Supervised and maintained underwater closed-circuit television equipment, digital computer equipment, and electronic navigation equipment.

Nuclear-powered Submarine   1988 - 1989

- Operated and maintained electrical generating and distribution equipment.
- Performed vibration analysis of rotating equipment.

EDUCATION/   B.S., Electronic Engineering Technology
TRAINING   University of California, Santa Barbara

USN Nuclear Power School
USN Nuclear Power Prototype
USN Electrician's Mate "A" School

JUDITH W. SWENSEN
4422 Kennet Avenue
Jubal, Tennessee 37232
615/555-4876

Summary of Qualifications

Experience in providing comprehensive environmental assistance to mining operations and exploration projects. Maintain an awareness of all federal environmental regulations to assess compliance of subsidiary companies. Conduct detailed environmental audits at mining and terminal locations.

Accomplishments

Solid Waste Disposal

As disposal methods analyst, charged with determining best disposal method at each subsidiary mine. Methods chosen are site-specific and depend on depth to groundwater, percentage and types of heavy metals present in the coal ash, and column leachate test results.

Investigated and designed economical solid and hazardous waste disposal options for subsidiary companies.

Mine Drainage Treatment

Assisted subsidiaries with effective economical methods of controlling acid mine drainage from coal refuse piles and ensuring reclamation success.

Conducted research with Tennessee State University to determine methods of refuse pretreatment to eliminate future AMD and have successfully installed two systems.

Employment History

Senior Development Specialist, 1997 to present
Smoky Mountain Mining Company, Memphis, TN

Graduate Assistant/Lab Technician, 1994 to 1997
University of Tennessee, Chemical Engineering Department

page 1 of 2

Judith W. Swenson
page 2 of 2

Education

University of Tennessee
Ph.D. in Chemical Engineering, 1997
B.S. in Chemistry and Physics, 1994

Affiliations

Scientific Society of America
Women in Environmental Engineering
Professional Engineering Association

References provided on request

Samantha T. Shavers
15 E. Green Street, #333
Richmond, VA 18978

804/555-3903

Overview

Experienced technical writer capable of producing quality

- product proposals
- advertising and catalog copy
- technical manuals
- scientific research proposals
- documentation for software systems
- environmental impact statements

Clients

- Jenkins Manufacturing, Richmond, VA
- Cooper Technical Publications, Atlanta, GA
- Electronic Design, Inc., Richmond, VA
- Jones & Wright Environmental, Inc., Washington, D.C.
- Software Solutions, Inc., Atlanta, GA

Credentials

B.A. in English, University of Wisconsin
June 1996
Minor: Computer Science

Member, Society for Technical Communications

References

References and writing samples available

STANLEY TRUMBULL
3 S. Sioux Trail
Ottawa, Ontario, Canada K1P 5N2
613/555-1782

OBJECTIVE: A career in the field of anthropology

EDUCATION: UNIVERSITY OF OTTAWA, Ontario, Canada
                  B.A. in Anthropology, expected June 2000

HONORS: Dean's List, 1999
                  Phillips Anthropology Award, 1999

EMPLOYMENT
HISTORY: OTTAWA UNIVERSITY, Ontario, Canada
                  Department of Animal Behavior
                  Research Assistant, 9/9 - Present
                  Input data for animal behavior studies.
                  Maintain lab equipment.
                  Monitor animals and recorded data.

                  OTTAWA UNIVERSITY, Ontario, Canada
                  Admissions Office
                  Student Assistant, 9/98 - 4/99
                  Conducted campus tours.
                  Processed applications.
                  Assisted in student recruitment and general
                  public relations.

                  PARKER & PARKER, Detroit, MI
                  Office Assistant, 6/98 - 9/98
                  Handled data entry, processing orders, phones.

ACTIVITIES: Anthropology Club, 1998 - Present
                  Student Government Representative, Fall 1998

References Available

GREG GOLD
1661 Corn Row Drive
Cedar Rapids, IA 53309

319/555-2909 (Home)
319/555-8888 (Work)

JOB OBJECTIVE: Engineering Technician/Camera Operator

OVERVIEW: Experience with all camera operations for film and video. Skills include studio lighting, set design, film editing, dubbing, gaffing, audio-video switching, mixing, and technical troubleshooting.

EXPERIENCE: WCED-TV, Cedar Rapids, IA
Engineering Assistant, 7/99 - present

Drawbridge Productions, Des Moines, IA
Assistant Camera Operator, Summer 1998

WWOR Radio, Jackson, MS
Engineer, 9/97 - 6/98

EDUCATION: Jackson University, Jackson, MS
B.A. in Communication Arts, June 1999

REFERENCES: Available on request

GRADY BISHOP

127 Golf Street
Apartment #8B
West Lafayette, Indiana 47906

Home: (317) 555 - 9876   Work: (317) 555 - 9854

OBJECTIVE: To find a challenging position in the aerospace industry that would
               utilize my engineering skills.

EDUCATION: Embry-Riddle Aeronautical University, Daytona Beach, FL
               Degree: Bachelor of Science in Aerospace Engineering

EXPERIENCE:

Jan. 97 to   Aeroflight International, Lafayette, IN
Present   Title: Strength Engineer

                Responsible for detailed stress analysis for engine components.
                Hand and finite element methods are utilized to examine the
                structural adequacy of various components of the fan and core
                thrust reversers (fixed & translating parts), the composite inlet and
                accessory compartment doors, and the fixed fan duct. Analysis
                includes static, thermal, and pressure loads in conformance with
                military standards. Interface closely with the Design group during
                the preliminary release phase to accelerate and optimize drawings.

June 93 to   Commercial Aircraft Program
Jan. 97   Title: Value Engineer

                Assigned to a training program to interface with manufacturing.
                Goal of the project was to discover fabrication techniques and
                difficulties and to improve channels of communication between
                engineers and manufacturing personnel. Concepts of value
                engineering were used on selected intensive and repeating
                problems. The project chosen was the air-conditioning system for
                the mid-size commercial aircraft. Cost savings realized through this
                program were considerable.

page 1 of 2

Grady Bishop
page 2 of 2

EXPERIENCE (cont.):

May 90 to June 93	Elite Aerospace Systems, Desert Palms, CA Title: Stress Engineer

Responsible for engineering analysis and structural substantiation on modifications for various commercial aircraft. Worked closely with FAA Designated Engineer Reps (DER) in design support work.

COMPUTER
EXPERIENCE: FEM programs such as NASTRAN, PATRAN and PIPELINE on both VAX and IBM.

REFERENCES AVAILABLE

CURRICULUM VITAE

SCOTT M. FRANK, M.D.

PERSONAL DATA

Birthplace	Dayton, OH
Citizenship	U.S.A.
U.S. Social Security	555-55-5555

Home Address	983 Crestview Drive
	Osceola, IN 46544
	(219) 555-9872

EDUCATION

Year	Degree	Institution
1982	B.S.	University of Cincinnati
1987	M.D.	University of Cincinnati, College of Medicine

POSTGRADUATE TRAINING

Year	Position	Institution
1987 - 1990	Residency	East Virginia Graduate School of Medicine, Norfolk, VA
1990 - 1992	Fellow, Nephrology	University of Iowa Hospitals and Clinics, Troy, IA

PROFESSIONAL EXPERIENCE

Year	Position	Institution
1992 - Present	Private Practice in Nephrology, Dialysis, and Transplantation	Scott Frank, M.D. Health Services 8978 Foxworth, Suite 555 Osceola, IN 46244

Page 1 of 2

Scott M. Frank
Page 2 of 2

APPOINTMENTS

Medical Director: Dialysis Transplantation September 1998	Osceola Medical Center Osceola, IN
Clinical Assistant Professor January 1998	Department of Internal Medicine University of Osceola Osceola, IN

COMMITTEES

Pharmacy and Therapeutics, Osceola Medical Center
Member, 1995 - 1998
Chair, 1998 - Present

Institutional Review Board
Osceola Medical Center
Member, 1996 - 1998

Capital Equipment
Osceola Medical Center
Member, 1998 - Present

CERTIFICATIONS AND LICENSURE

Certification

American Board of Internal Medicine - 3/26/94 - #555555
Nephrology, American Board Internal Medicine - 11/11/92 - #555555

Licensure (current)

Indiana - 7/2/94 - #555555

Frederick P. Sroblewski
3456 Muscatel Ave
Tucson, Arizona 85718
(602) 555-6543

BACKGROUND:

Award-winning technical educator and computer support professional with international experience. Able to make an outstanding contribution to your organization in areas of:

- **Management Information Systems**
- **Technical Education**
- **Technical Writing**
- **Computer Systems Support**

SELECTED QUALIFICATIONS:

Technical Trainer - Achieved outstanding recognition as an educator. Selected and supervised staff of instructors.

Creator and Developer - Researched, created, packaged, and implemented training programs never before taught in areas of quality management, computer hardware, computer software, and concepts from ideas/needs to systems application.

Technical Writer - Track record of writing program manuals, self-taught programs, and on-the-job training manuals. Served as resource to writing staffs to evaluate and rewrite material for practical instruction.

Increased Revenues - Because of personal excellence in providing training for both foreign and domestic clients, additional training services were purchased in multiple modules of $100K plus.

Computer Support Specialist - Comprehensive experience regarding computer hardware troubleshooting and repair of equipment, maintenance of integrated systems from mainframe to micro, including peripherals, used throughout the industry.

SELECTED CAREER ACHIEVEMENTS:

TRAINING SPECIALIST
(John Zink Company) (1998 - present)

- Create, develop, and conduct training for both employees and customers in formal and informal settings.

Page 1 of 2

SELECTED CAREER ACHIEVEMENTS (cont.):
- Evaluate courses to determine quality of content and format as well as recommend selection of appropriate staff trainers.
- Was directly used as instructor with clients in ten countries as well as in the United States.

TECHNICAL TRAINER
(U.S. Army) (1990 - 1998)

- Responsible for training theory and application of electronics to nontechnical personnel, resulting in exceptional number of participants being assigned to technical responsibilities.
- Responsible for providing total systems support under all conditions for assigned duty, with a battlefield support system.

EDUCATION:

U.S. Army Electronic Training, 1991 - 1994

- Achieved highest score of any student during school's 15-year history.

SPECIAL TRAINING AND EXPERTISE:

- Special purpose computers
- Super computer integrated systems
- Mainframes through micros
- Peripheral equipment
- Personal computer applications

REFERENCES:

Furnished upon request.

Ernest Martin
4366 South Street
Detroit, MI 48062
(616) 555-9698

Career Goal: To obtain a position teaching dental hygiene.

Education: Temple University, Philadelphia, PA
M.S. Dental Hygiene, 1998

Western Michigan University, Kalamazoo, MI
B.S. Dental Hygiene, June 1996

Work Experience: June 1998 - Present

Western Michigan University, Kalamazoo, MI
Instructor of Dental Hygiene Classes: oral anatomy, periodontology, and physiology.

September 1996 - May 1998

Western Michigan University, Kalamazoo, MI
Graduate Assistant
Duties: teaching section in periodontology and physiology, grading assignments and quizzes, and recording attendance for lecture periods.

References: Available upon request.

EDUARDO LOPEZ

6 E. Columbus Drive
College Park, MD 20740
410-555-3938

Goal: Research technician position that allows me to use my training in physics.

Education: University of Maryland, College Park, MD
B.S., Physics, June 1999

Relevant Coursework

Plasma Physics
Medical Instrumentation
Statistics
Research Methodology

Honors

Dean's List
Sigma Pi Sigma, Physics Honor Society

Experience: University of Maryland, Physics Department
Research Associate
June 1999 to Present

Conduct literature research and create literature studies to support work of department. Record and analyze research data. Contribute to technical reports and publications.

Huntington Burroughs Pharmaceutical, Inc.
Student Intern
Summer 1998

Assisted senior researcher with data input, statistical analysis, and computer model development.

References: Available on Request

EDWARD J. FISHER
1456 Burlington Ave.
Cincinnati, Ohio 45642
(513) 555-8976

CAREER OBJECTIVE

A cost-effective performer with a proven record of accomplishment, my career objective is to utilize my management, marketing, and computer service experience to make an immediate contribution as a member of a professional management team.

OPERATING SYSTEMS EXPERTISE

MVSIXA, CICS, Vtam, CA/I, CA/7, NCP, SURPA, VPS, ACF2, Omegamon

HARDWARE

IBM 309X, 308X, IBM 4331, StorageTek 4400 ACS

CAREER SUMMARY

OPERATIONS MANAGER: 1994 - Present
Genair Corporation, Cincinnati, OH

Achievements:

Directed implementation of new data center and hired and trained operations and network personnel. Brought on-line ten months under scheduling.

Reduced printing costs by 50% resulting in an annual savings of $1.5M by managing a project team through the analysis, design, development, and implementation of new printing systems and procedures.

Installed corporate telecommunication systems, including PBX'S, key systems, and national contract with a major carrier, resulting in a savings of over $1M.

Planned and monitored annual operating budget, supervising technical staff consisting of 25 supervisors, analysts, operators, and remote schedulers.

Page 1 of 2

EDWARD J. FISHER
Page 2 of 2

CAREER SUMMARY (cont.)

OPERATIONS MANAGER: 1985 to 1994
Information Services Agency, Dayton, OH

Achievements:

Initiated automated problem resolution system resulting in reduction of recurring problems and elimination of tedious manual system.

Developed position titles and pay scales that resulted in identifiable career paths for operations personnel.

EDUCATION

Computer Science Degree from Rose-Hulman Institute of Technology, Terre Haute, Indiana

Master in Business from University of Notre Dame, South Bend, Indiana

PROFESSIONAL TRAINING

Managing Data Processing/IBM

Data Processing Operations Management/IBM

Turning Telephone Costs into Profits/University of Notre Dame

PROFESSIONAL AFFILIATIONS

ITUA - Indiana Telecommunications User Association

AFCOM - Association for Computer Operations Managers

REFERENCES AVAILABLE UPON REQUEST

■ ─ ■ ─ ■ ─ ■ ─ ■ ─ ■ ─ ■ ─ ■ ─ ■ ─ ■ ─ ■ ─ ■ ─ ■

EARLE F. ABLE
6255 Maple Drive
Chicago, IL 60636
──────
Telephone (312) 555-4948
Fax (312) 555-4897

OBJECTIVE	To attain a position as a designer drafter with a highly aggressive, goal-oriented architectural engineering firm.
SKILLS	Design and detailing of commercial, mechanical, electrical, and plumbing systems
	Architectural plan and details and site layout

CAREER EXPERIENCE

May 1997 to Present	ENGINEERING DEVELOPMENT DESIGN Position: Designer Drafter
	Duties: Design development of mechanical, electrical, and plumbing systems within commercial projects. Produce final bid documents on multiple medias and AutoCAD software. Develop construction details for architectural and engineering concepts. Also responsible for pictorial sections used in development and layout.
EDUCATION	B.S. Mechanical Engineering Technology University of Illinois, Champaign, 1997
	A.S. Mechanical Design and Drafting Technology Indianapolis Community College, 1995
REFERENCES	Furnished upon request

EDDIE FISHER

Campus Address	Home Address
8223 Green Street	93 West 4th Street
Pasadena, CA 91125	Long Beach, CA 90808
(818) 555-7879	(213) 555-9876

EDUCATION

June 1997	B.S. in Civil Engineering	California Institute of Technology
	G.P.A. 3.67	Pasadena, CA

WORK EXPERIENCE

August 1997 -		
Present	*Project Engineer*	Welding Corp., Culver City, IA

Main projects consisted of bridge re-decking and finalizing the construction of a cut-and-cover tunnel. Responsibilities included:
- Coordination of all subcontractors and suppliers with JBC and DOT
- Scheduling weekly quantity surveys
- Estimating weekly budget reports
- Interpretation of drawings and specifications

January 1990 -		
August 1997	*Assistant Project Manager*	Grote Construction, Duluth, MN

Responsibilities included:
- Estimating costs
- Quantity take-off
- Crew sizing
- Scheduling
- Job cost control subcontractor
- Quantity surveys
- Design and determination of construction methods

HONORS AND AFFILIATIONS

Chi Epsilon (XE) National Civil Engineering Honor Society
Dean's Honor List, College of Engineering
American Society of Civil Engineers (ASCE)

COMPUTER SKILLS

Proficient with Macintosh and IBM computers, associated software programs. BASIC and PASCAL programming, STRUDL, experienced in the NMSU mainframe.

ELIZABETH ENGLE
2316 King Street
Richardson, TX 75080
972/555-2552

GOAL

Petroleum engineering position with small, independent oil exploration and production company.

EDUCATION

University of Texas at Dallas
B.S. in Petroleum Engineering, expected May 2000

COURSEWORK

Petroleum Engineering Design
Rocks and Fluids
Reservoir Modeling
Reservoir Engineering
Secondary Recovery
Drilling Design & Production

WORK EXPERIENCE

UNIVERSITY OF TEXAS AT DALLAS
Lab Assistant/Physics Dept., 1998 - 2000

Assisted professors in the Physics Department with lab experiments and general office work.

MEMBERSHIPS

Society of Petroleum Engineers
Engineering Club

SPECIAL SKILLS

Working knowledge of Lotus 1-2-3 and Wordstar

REFERENCES AVAILABLE

Evelyn Moore
4366 South Street
Detroit, Michigan 48062
(616) 555-9698

Career Goal: To obtain a position as a secondary education instructor in the areas of Science and Computer Science.

Education: September 1997 to present

Western Michigan University, Kalamazoo, Michigan 49008
Secondary Education Curriculum
Biology Major, Computer Science Minor

September 1993 to June 1997

Littlefield Public School, Albert, Michigan 48076
Graduated salutatorian, June 1997

Work Experience: February 1999 to present

McDonald's, 39 King Drive, Kalamazoo, Michigan 49008
Phone (616) 555-5137
Swing Manager. Duties: cash audits, deposits, quality control of product, customer relations, supervision of employees, inventory and ordering of supplies, maintenance of restaurant appearance, register operations, and associated paperwork.

May 2000 - Present

Computer Science Department, Western Michigan University, Kalamazoo, Michigan 49008
Phone (616) 555-4620
Computer Operator. Duties: software inventory and evaluation, programming, entering and updating files, journal photocopying, and article synopsis.

References: Available upon request

Edward Dumphy
566 Perry Blvd.
Altus, Oklahoma 74170
(918) 555-7809
Fax (918) 555-7778

OBJECTIVE: FIELD ENGINEER

To manage heavy highway construction projects.

EDUCATION

Purdue University
Bachelor of Science, Construction Management, May 1996

RELATED COURSEWORK

Temporary Structures	Electrical/Mechanical Systems
Construction Equipment	Heavy Construction Estimating
Soils in Construction	Legal Aspects in Construction

EXPERIENCE

WILLIBROS BUTLER ENGINEERS, INC.
Altus, Oklahoma
Project Engineer
6/99 to Present

Interstate 465 Widening Project; contract value $55 million. Responsible for internal and subcontractor payletter quality, subcontractor negotiations with the State of Oklahoma Department of Transportation, and subcontractor scheduling. Produced weekly and monthly cost/quantity reports. Processed extra work bills and time cards. Managed punch list and construction crews. Assisted project superintendent in selling of job.

ROCK WARE CONSTRUCTION COMPANY
Oklahoma City, Oklahoma
Assistant Operations Manager
5/97 to 6/99

Responsible for residential demolition, installation of concrete footings, drywall, and painting. Performed concrete quantity takeoffs and job setup/preplanning. Practical understanding of construction problems. Developed teamwork skills.

page 1 of 2

Edward Dumphy
page 2 of 2

EXPERIENCE (cont.)

WINDSOR SHIPPING COMPANY
Windsor, Ontario, Canada
Shipping and Receiving Coordinator
Summers 1992, 1993

Responsible for all phases of shipping and receiving. Packaged merchandise and performed computerized inventory control. Processed purchase orders. Served as forklift operator, truck driver, and in various clerical capacities. Developed understanding of contractual relationships.

REFERENCES AVAILABLE UPON REQUEST

David T. Sanchez
10001 W. Edina Ave.
Edina, MN 53989
612/555-5453

STRENGTHS:

• Excellent communication and people skills

• Strong photographic and processing skills

• Academic and hands-on training in commercial art

• Computer literate, with working knowledge of QuarkXPress and PageMaker

EDUCATION:

University of Minnesota, St. Paul, MN
B.A. in Commercial Art, expected May 2000

WORK EXPERIENCE:

Minneapolis Magazine, Minneapolis, MN
Commercial Artist, Summers 1997 - present

University of Minnesota, St. Paul, MN
Designer, University Publications, 1999

University of Minnesota, St. Paul, MN
Photographer, Student Gazette, 1998 - 1999

WORKSHOPS:

Website Design Seminar, University of Minnesota, 1999
Illustration Workshop, Art Institute of Chicago, 1998
Midwest Design Seminar, Northern Illinois University, 1997

REFERENCES AVAILABLE

MARY SMITH
543 Hillside Drive
Palo Alto, CA 94304

(415) 555-9876

OBJECTIVE:	To apply rigorous quantitative methods and models to strategic planning issues
EDUCATION:	Columbia School of Engineering and Applied Science B.S. in Operations Research Concentration: Computer Science Degree Expected: June 2000 Courses Taken: Mathematical Programming Data Structures and Algorithms Accounting and Finance Production-Inventory Planning and Control
COMPUTER SKILLS:	Languages: BASIC, C, FORTRAN, PASCAL Hardware: DEC-29, Sun Work Station, IBM, and Macintosh Software: Lotus 1-2-3, various business software systems Operating System: Tops-20, UNIX, DOS
EXPERIENCE:	Research Assistant to Professor Samuel Silvers Department of Operations Research Columbia University, New York, NY • Developed and implemented performance analysis of scheduling • Designed scheduling system for university laboratories
REFERENCES:	Available

DAN LUI

17 Dinge Road
Terre Haute, IN 52211
317/555-1331 (Home)
317/555-2339 (Office)

OBJECTIVE: A position in the field of Electrical Engineering with an emphasis on
                        aviation electronic systems.

EDUCATION: B.S. in Electrical Engineering, May 1999
                        Rose-Hulman Institute of Technology, Terre Haute, IN
                        G.P.A. 3.75
                        Graduated with Honors

WORK EXPERIENCE: *C & S Industrial Design Consultants, Richardson, TX*
                        Summer Intern, 1998
                        Assisted in research and development department of aviation
                        electronics firm. Input data, typed performance specifications reports,
                        calibrated lasers, and maintained test equipment.

                        Rose-Hulman Institute of Technology, Terre Haute, IN
                        Assistant to the Director, Financial Aid, 1997 - 1998
                        Processed applications. Handled general office duties.

ACTIVITIES: President of Student Chapter of Institute of Electrical and Electronics
                        Engineers
                        Peer Advisor, Engineering Department

REFERENCES: Available upon request

DARNELL GRANT
800 York Ave. South
Minneapolis, MN 55410
(612) 555-7908

EDUCATION

B.S. Mechanical Engineering Technology, Rose-Hulman Institute of Technology, 1996
(GPA: 5.75/6.00)

EMPLOYMENT

7/98 to present: TECHNASTAR DESIGN DEVELOPMENT

POSITION Development and Design Engineer (7/98 to present)

   Responsible for the design and applications of product components for automotive plastic body panels. This includes recommending design changes and testing product for reliability and durability.

Project Engineer (6/98 to 7/98)

   As a project engineer I was involved in manufacturing procedures, processing improvements, and troubleshooting production problems for tooling. I managed maintenance activities and vendor contracts for production tooling.

TECHNICAL SKILLS

Computer - AutoCAD, BASIC, Lotus, VP-Expert

AFFILIATIONS

Member of Society of Automotive Engineers
Member of Society of Plastics Engineers

UNIVERSITY OF KENTUCKY
CAREER PLANNING AND PLACEMENT CENTER

LEXINGTON, KENTUCKY 40506

NAME:	Alicia Curtiss	
PRESENT	9867 High Drive	TELEPHONE:
ADDRESS:	Lexington, KY 40506	(415) 555-8712

OBJECTIVE: A position in research and development with a company interested in wide applications of polymeric materials. Possibility for move into management preferred.

EDUCATION: UNIVERSITY OF KENTUCKY, Lexington, KY
M.S., Chemical Engineering, 1998

LOUISIANA TECH UNIVERSITY, Ruston, Louisiana
B.S., Chemical Engineering, 1996

EXPERIENCE:

7/98 - present   University of Kentucky, Dept. of Chemical Engineering, Lexington, KY
Research Assistant, laboratory of Dr. S. Silverberg:
Study structures formed by diblock copolymers in a solvent selective for one block. Use light, x-ray, and neutron scattering to determine micelle structure as a function of solution conditions. Compared the spherical micelles to structures predicted for multi-armed star polymers.

5/92 - 9/96   University of Kentucky, Dept. of Chemical Engineering, Lexington, KY
Research Assistant:
Studied properties of monolayer and multilayer films of alkanoic acids and alkylsiloxanes on solid surfaces.

REFERENCES: On Request

ADAM CANTOR

Work Address:	Department of Chemistry University of Vermont Burlington, VT 16901 (802) 555-9811
Permanent Address:	2141 Rock Street Alameda, CA 94501 (802) 555-1123

OBJECTIVE

To obtain a position as a senior research and development chemist in the fields of polymer or physical chemistry.

EDUCATION

Ph.D., University of Vermont, 1999
in Physical Chemistry, Dynamic Light Scattering Study of Ternary Polymer Solutions

B.S., Middlebury College, 1991
in Chemistry
Minors: Mathematics, Physics

EXPERIENCE

6/92 to 8/99
University of Vermont, Department of Chemistry

Research Assistant. Studied semi-dilute poly (n-alkyl isocyanate) solutions containing a linear polystyrene probe polymer. Examined concentration, molecular weight, and temperature dependences of the ternary solutions. Performed dynamic and static light scattering measurements. Characterized solutions using FTIR, UV, viscometry, and differential refractometry. Assisted in design, assembly, and maintenance of experimental instruments. Administered laboratory computer systems (UNIX, VMS, Macintosh, MS/DOS).

2/93 to 1/98
University of Vermont, Department of Chemistry

Health and Safety Representative. Implemented laboratory safety measures, prepared chemical inventories, and provided personal safety instruction. Aided in development of departmental safety film.

Page 1 of 2

Adam Cantor
Page 2 of 2

EXPERIENCE (cont.)   6/90 to 12/90
                     Middlebury College, Department of Chemistry

                     Teaching Assistant for Physical Chemistry, Laboratory
                     (Head TA and Course TA), First-year Chemistry, Organic Chemistry

HONORS   Member, Eta Kappa Nu Fraternity
                     Recipient, Chemical Engineering Scholarship of America
                     National Merit Scholar

REFERENCES   Furnished on Request

ALLEGHENY COLLEGE
CAREER PLANNING AND PLACEMENT CENTER
MEADVILLE, PENNSYLVANIA 16335

Name: Allen Day
Address: 88 State Street
           Carmel, IN 46032

Phone: (317) 555-3175

OBJECTIVE: To obtain a position in information systems, software
             design/development, or related area utilizing computer programming
             language skills.

EDUCATION: Allegheny College, Meadville, PA
             B.S., Computer Science
             Graduation Date: June 1999

EXPERIENCE:

Summer 1998   City of Reading, Reading PA, *Management Information Systems Intern:*
              Duties included personal computer assembly and setup (hardware &
              software installation) as well as system troubleshooting. Involved
              significant user interaction and operating system knowledge. Worked
              on IBM PC AT's and XT's, HP Bectra PC's using MS DOS 3.3.

Summer 1997   Indiana University, Indianapolis, IN, *Student Programmer:* Developed
              application that aids in vision/perception research by performing linear
              transformations to bitmap images. Consultant to supervisor.

Summer 1996   Indiana University, Indianapolis, IN, *Research Programmer:* Developed
              an IBM application for desktop security and screen-saver.

SKILLS SUMMARY:

Computers: C, Pascal, Lisp. Also familiar with Ada, SmallTalk, Prolog, 68000
             Assembly Language. Procedural, Functional Object-oriented
             programming. LightSpeed, MDSenvironments.

REFERENCES: Available on Request

Allen Fisher
88 Waverly Road
Huntington, IN 46872
(317) 555-9876

OBJECTIVE:

A position in the architectural/engineering/interiors field with emphasis in the CADD environment.

EXPERIENCE:

4/97 - Present: Tad Technical Services (Dow Chemical Company) as a CADD drafter.

Working on a team, my duties are revising drawings to "as built" status. Working with Digital 3100 computer running Autotrol Release 8.2 software on VAXNMS V5.3-1 operating system.

1/97 - 4/97: Manpower Technical Services (DBC Architects, Inc.) as a CADD drafter.

Duties included drafting up revisions on floor plans, reflected ceiling plans, schedules, and details on various projects. Worked with a Compaq 386 computer running Autocad Rclease 10 software.

7/96 - 1/97: SAL, Inc. as a CADD drafter.

Duties included space planning, details, and product design of various fast food restaurants. Worked with a Compaq 386 (clone) computer running Cadmaza software. Some exposure to Unix on Sun Sparcstation.

9/95 - 6/96: Boeing Products, Inc. as a CADD drafter (Autocad).

10/94 - 8/95: West's Architects, Inc. as a CADD drafter (Autocad).

EDUCATION:

Syracuse University, NY
M.S., Operations Research

Stanford University, Stanford, CA
M.Sc., Statistics

Duke University, NC
B.Sc., Mathematics

Page 1 of 2

Allen Fisher
Page 2 of 2

SKILLS:

Knowledge of Basic, COBOL, working knowledge of FORTRAN, Pascal, Excel, Lotus 1-2-3

MEMBERSHIPS:

Operations Research Society of America
Society for Industrial and Applied Mathematics
American Association of Computer Professionals

REFERENCES:

A detailed list of professional references will be provided on request.

ALEXANDER HO

986 Parker Lane   Work (415) 555-2939
Walnut Creek, CA 94595   Home (415) 555-9875

SUMMARY

Intimate knowledge of microcomputer industry and applications software. More than ten years of broad international business experience with Fortune 500 corporations. Fluent in Mandarin Chinese, good command of Japanese and Thai. Successful accomplishments in:

Feasibility Studies   Office Automation

Multinational Manufacturing   Economic Recovery and Product
Analysis and Control   Positioning and Pricing

Strategic Planning and   Marketing Plans and Strategies
Competitive Analysis

EXPERIENCE

HOCORP INTERNATIONAL, INC., WALNUT CREEK, CA

1991 - Present **President**
             Founded consulting and marketing firm to evaluate business
             problems, determine software requirements, and develop
             microcomputer systems to meet clients' needs.

INTERCORP INC., LOS ANGELES, CA

1990 - 1991 **Manager, New Product Programs**
             Responsible for product evaluation, assessment of marketing
             potential, and development of product feasibility studies.

             Completed feasibility study on 9500 Electronic Printing System
             for the Pacific Rim area, which led to the development of a new
             market area.

Page 1 of 2

INTERCORP (cont.)

Evaluated marketing strategies for microcomputer products in open market countries.

1988 - 1990   **Manager, Field Pricing**
Developed, evaluated, and recommended strategic and tactical pricing actions enabling affiliates to exceed targeted profits.

Developed and implemented major account pricing strategy for Malaysia resulting in an increase on major accounts and a reduction in cancellations.

1984 - 1988   **Manager, Commercial Analysis**
Direct responsibility for long-range competitive forecast for group of 25 affiliates. Measured performance of current products, competitive practices, and identified risks and opportunities to business strategies.

EDUCATION

M.B.A.	1981	University of Georgia, Athens, GA
B.S.	1980	Notre Dame University, South Bend, IN Major: Industrial Administration
B.A.	1978	City College, Chicago Major: Marketing

JOHN J. ALLEN

Present Address   Permanent Address
765 5th Street   28 Octavia Terrace
Washington, D.C. 20016-8001   Washington, D.C. 20019
(202) 555-2213   (202) 555-9737

OBJECTIVE   Full-time position as a medical writer for a pharmaceutical
                 company, medical school, textbook publisher, or government
                 agency.

EDUCATION   George Washington University
                 Currently pursuing M.S. in technical writing with a
                 concentration in biology.

                 Whitman College
                 B.S. with Highest Distinction in English, May 1999

EXPERIENCE   Eli Lilly and Company, Indianapolis, IN
                 Summer 1998

                 SUMMER INTERN:
                 Analyzed new product data and prepared reports for in-house
                 use by sales staff. Interviewed researchers and prepared articles
                 for company publications.

                 Washington Post, Washington, D.C.
                 Summer 1997

                 SUMMER INTERN:
                 Wrote columns on health fads, fitness, and new drugs.

CREDENTIALS   A.M.W.A. certificates in pharmaceutical writing and editing
                 Member, American Medical Writers Association Editor, Whitman
                 College newspaper

REFERENCES   On request

CHARLES ANDAWA

4783 West Maple • Baltimore, MD 21218 • 410 555-2983

Objective   A position in chemical engineering with an environmental engineering organization.

Recent Experience

1993 - Present   Senior Project Manager, PPD, Inc., Baltimore, MD

Report to senior vice president of engineering. Supervise 32 employees. Direct, supervise, administer, and manage projects from inception to startup, including new chemical process equipment manufacturing. Assist sales department in reviewing the system process design, scheduling, engineering, and costs before final proposal is presented to client.

Conceive, initiate, and develop chemical formations for nontoxic solutions for use in oil recovery and recycling. Formulated empirical equations and design criteria for the system, which resulted in increasing company sales seven-fold over the last five years.

Train project engineers and project managers to design and manage projects.

1991 - 1993   Senior Project Engineer, Moreland Chemical, Annapolis, MD

Project experience included pulp liquor evaporation system operations, sand reclamation systems, waste wood utilization to manufacture charcoal, sewage sludge oxidation, waste oxidation, and heat recovery. Responsible for planning, scheduling, process design, and specifications.

Education   B.S., M.S., Chemical Engineering, Minnesota Technical University, Houghton, MN, 1984

References   Available by request.

Alicia Anderson
2411 White Street
Des Plaines, IL 60016

(847) 555-8368

SUMMARY

Experienced freelance technical writer seeking new clients.

SKILLS

- Development of product proposals
- Content and copyediting of technical documents
- Revision and fact checking of technical manuals
- Research projects
- Documentation for software products
- Creation of revision reports for technical engineers

COMPUTER SCIENCE BACKGROUND

COBOL	PL/1	DB2	BASIC
RPG III	C	C Plus	BAL
FileAid	Ada	CAD/CAM	

OTHER CREDENTIALS

University of Wisconsin
B.S. in Computer Science, June 1999

Member, Society for Technical Communication

REFERENCES

Writing samples, client list, and references on request.

Jason Baxter
93 West 4th Street
Long Beach, CA 90808
(213) 555-9876

EDUCATION

June 1996:   Bachelor of Science in Cytotechnology
California Institute of Technology
Pasadena, California
GPA 3.65
Relevant Courses: bacteriology, physiology, anatomy,
histology, embryology, zoology, genetics, chemistry,
and computer classes.

WORK EXPERIENCE

June 1998 - present: CYTOTECHNOLOGIST
Circle Center Research Laboratory
Culver City, CA

Duties: To identify cell specimens that are collected by
fine-needle aspiration and report findings to the
pathologist. Use computers to measure cells, a new
technique that is being developed at Circle Center
Research Laboratory.

CERTIFICATION

The International Academy of Cytology

National Certification Agency for Medical Laboratory
Personnel

HONORS

Dean's Honor List - six semesters

REFERENCES

Available upon request

BRUCE CATT
6543 MAPLE STREET
GREENWOOD, INDIANA 46142
PHONE 317-555-1123

EDUCATION

- B.S. Geology; Miami University - Oxford, Ohio - 1998
- Graduate Studies in Geology; Oberlin College - Oberlin, Ohio
- Forestry and Natural Resources; Ohio Northern University - Ada, Ohio
- OSHA 29 CFR 1910.120 training

PROFESSIONAL EXPERIENCE

Project Geologist

Computers & Structures, Inc., 4/98 - present

- Responsibilities include managing over 200 environmental assessments for properties undergoing acquisition or refinancing.
- Supervision and documentation of underground storage tank removal and closure and subsequent contaminated soil remediation.
- Responsible for remedial investigations including subsurface investigations to delineate the extent of soil contamination, design and installation of monitoring well systems, groundwater sampling and analysis, soil gas surveys, and geophysical studies.
- Design of soil venting systems, groundwater recovery/treatment systems, and bioremediation programs.
- Project management responsibilities include proposals, drill scheduling, material purchasing, invoicing, and client development.

AFFILIATIONS

- Ohio Academy of Science
- Geological Society of America

References are available and will be furnished upon request.

Name:	Maria Black
Address:	1419 Cedar Drive Dayton, OH 45226
Phone:	(513) 555-8754
Qualifications:	Bachelor of Science in Pharmacy, 1996
School of Pharmacy:	Dayton University
Special Award:	Merrell Dow Dayton School of Pharmacy's Annual Award for Excellence, 1999

Previous Experience:	Hooks Pharmacy Dayton, OH	Summer Student 10 weeks, 1999
	Royal Hospital Dayton, OH	Summer Student 8 weeks, 1998
	Ohio Drug Dayton, OH	Saturday Staff 9/94 - 1/99

Present Position:	Pharmacy Graduate Intern Program Dayton Community Hospital Dayton, OH
Interests:	My main interest is in clinical pharmacy.
	During my intern year, I have attended Dayton University evening classes on clinical pharmacy and an Ohio State University course in ambulance first aid.
References:	Available on request.

BRADLEY Q. TRAPP

1726 Willow Springs Walk, Blue Springs, MD 64015

(816) 555-9997

OBJECTIVE

Senior Executive - Construction/Engineering

SUMMARY

Executive with diversified construction contract management achievements in a variety of industrial, refinery, petrochemical, and power generation projects. Demonstrated ability to contribute to profitable operation and growth in accordance with short- and long-term goals. A leader with innovative, analytical, and communication skills, with dynamic results in cost-sensitive critical processes and special projects.

PROFESSIONAL EXPERIENCE

LAMBERT SKY SUPPLY, INC.   Blue Springs, MD   1995 - Present

Divisional Vice-President Sales
Complete profit and loss and operational responsibility for division of this equipment leasing company.

- Created and implemented strategic business plan for company operations throughout west central United States (20 states), resulting in the development of 50 new accounts.

- Increased utilization and occupancy rate of company properties from 50% to 95% during five-year period.

- Increased overall revenues by 25% from 1997 - 1999.

REASONERS, INC.   White Plains, IA   1993 - 1995

Executive Manager
Developed a professional project management program to utilize existing company resources and provide for diversification. The program outlined in depth a series of options for owners' use in their building program, from conceptual through start-up phase, providing for increased project utilization and reduced building cost.

Page 1 of 2

PROFESSIONAL EXPERIENCE (cont.)

PUBLIC SERVICE IOWA   Cedar Rapids, IA   1989 - 1992

Vice President and Director
Complete operational responsibility for this construction and engineering company.

- Supervised engineering, estimating, and construction of many multi-million dollar industrial projects.

- Directed overall operations to effect a three-year growth by 100%, to a total of 50 million dollars.

Vice President Construction
Developed and implemented administrative and financial controls to effect significant annual savings for construction projects.

WESTERN SYSTEMS   Tempe, AZ   1986 - 1989

Manager of Construction
Represented this construction company in a joint venture of power plant construction.

- Directed construction of six 500 megawatt power plants throughout the Midwest.

- Created and implemented cost reduction actions for plant construction that resulted in a 25% reduction per megawatt cost as compared to the national average.

EDUCATION

B.S. Civil Engineering - Michigan State University - 1987

References Available

Charles D. Stiles
8765 South East Street
Ada, OH 45810
(419) 555-9876

CAREER OBJECTIVE

Position that requires technical knowledge in the areas of design, testing, and reliability of mechanical and electrical systems in order to produce a quality product.

EDUCATION

Ohio Northern University, Ada, OH
B.S. in Mechanical Engineering Technology, 1996

Ohio Northern University
Have completed 21 hours of electronics and 15 hours of computer programming.

WORK EXPERIENCE

Shepherd Engineering - Tulsa, OK
1998 to Present
Responsibilities include:

- Component designs

- Thermoset and thermoplastic molding

- Tooling evaluation

- Assembly line setups

- Adhesive development

- Robot feasibilities

- Supplier contacts

Ford Motor Company - Detroit, MI
1996 to 1998
Responsibilities included:

- Traveling to various engineering facilities to develop tests

- Setting up inventory systems

- Maintaining budget

- Supervising laboratory technicians

- Publishing testing manuals and reports

Christopher Knight
1700 W. Armadillo
San Diego, CA 90087
619/555-9000 (work)
619/555-2839 (home)

OBJECTIVE: To obtain a position as Vice-President of Public Relations with an
               aeronautical corporation.

AREAS OF EXPERIENCE:

Marketing Development

• Initiated and supervised sales programs for aircraft distributors
selling aircraft to businesses throughout the western United States.

• Managed accounts with a profit range of $100,000 to $1,000,000,
including Dow Chemical, Landston Steel, Mercury Company,
Berkeley Metallurgical, and Ford Motor Company.

• Demonstrated to customer companies how to use aircraft to
coordinate and consolidate expanding facilities.

• Introduced and expanded use of aircraft for musical tours.

Public Relations

• Handled all levels of sales promotion, corporate public relations,
and training of industry on company use of aircraft.

• Managed promotions including personal presentations, radio and
TV broadcasts, news stories, and magazine features.

Pilot Training

• Taught primary, secondary, and instrument flight in single and multi-
engine aircraft.

Page 1 of 2

Christopher Knight-Page 2 of 2

EMPLOYMENT HISTORY: Hughes Aircraft, Inc., San Diego, CA
Sales Manager and Chief Pilot
1998 to Present

Boeing Corporation, Kansas City, MO
Assistant Manager of Promotion
1989 to 1998

American Airlines, Dallas, TX
Pilot
1982 to 1989

United States Air Force, Houston, TX
Flight Instructor
1980 to 1982

PROFESSIONAL LICENSE: Airline Transport Rating 14352-60
Single, Multi-Engine Land
Flight Instructor - Instrument

EDUCATION: University of Texas, Houston, TX
B.A. in History, 1961

MILITARY SERVICE: United States Air Force
1977 to 1982

REFERENCES: Available on request

DIANNE BRONOWSKI
◆ ◆ ◆ ◆ ◆ ◆ ◆ ◆ ◆ ◆
3355 Brookshire Parkway • Chicago, IL 60636 • Telephone (312) 555-4948 • Fax (312) 555-4897

OBJECTIVE: I hope to utilize my communication, problem-solving, and computer
               skills in an entry-level position with opportunities for advancement.

EDUCATION: ITT Business Institute, Associate Degree in Business, June 1999

EXPERIENCE:

5/99 to Present **Office Assistant**   **SERVICE SOFTWARE, INC.**

               Assist software design specialists. General clerical duties, including
               answering phone and creating correspondence. Most recent project
               is assisting with development of user manuals. Test manuals and
               provide notes to software designers so that they can implement
               changes.

6/97 to 5/99   **Executive Secretary**   **NEW WORLD PACKAGING**

               Responsible for clerical and receptionist duties for packaging firm.
               Maintained all office files and records. Produced correspondence.
               Directed all incoming calls. Provided basic customer service.

SKILLS: WordPerfect 6.1, Excel
               Typing Speed of 70 w.p.m.
               Ten-key Calculator by touch
               Some knowledge of Spanish

                         REFERENCES AVAILABLE

Dianne Hernandez
3125 Cool Creek Drive
Carmel, IN 46032
317-555-9406

OBJECTIVE   To obtain a position as an engineer with the opportunity to apply my knowledge of digital circuit design, programmable controllers, and microprocessors.

EMPLOYMENT

July 1998 to   Allied Wholesale Electrical Supply, Inc.
Present   Indianapolis, IN
                Systems Engineer

                Responsibilities include:

                • Resolving computer problems

                • Keeping inventory

                • Working with programmable controllers

                • Working with CAD

                • Analyzing change requests

                • Writing troubleshooting documentation

August 1994 to   Webber Engineering
June 1998   Carmel, IN
                  Die Detailer

                  Responsibilities included:

                  • Drawing and dimensioning die details

                  • Making engineering changes to die drawings

                  • Running blueprints

EDUCATION   Lawrence Institute of Technology, Southfield, MI
               B.S. Electrical Engineering, 1994

               Passed Professional Engineering Exam, June 1994

               References Will Be Provided Upon Request.

Chapter 6

Sample Cover Letters

Lee King
300 Lake Shore Drive, #22
Chicago, IL 60603

June 15, 20__

ENVIRON Management
300 Butterfield Road
Oak Brook, IL 60521

Dear Personnel Director:

I am seeking a position with your firm that utilizes my training in the environmental field. I am eager to work for a firm that is implementing the latest environmental advancements.

I earned a Bachelor of Science degree in Environmental Health Science from Indiana University. Through two paid internships with county health departments, I have participated in health and safety training, regulatory compliance, and on-site inspections. In addition, I am an experienced technical writer and am proficient with several computer programs, including Excel and WordPerfect. During college, I was a member of several organizations including Sigma Alpha Epsilon fraternity, the Environmental Health Association, and the Bicycling Club. I have developed leadership and team working skills through my work and college experiences.

I would be delighted to meet with you at your convenience to discuss career opportunities with your firm. I can be contacted by telephone at (312) 555-8961.

Sincerely,

Lee King

enclosure: resume

JASON DEAN
489 Sutton Street
New York, NY 10028
(212) 555-9085

May 3, 20__

Ms. Linda Appleton
Overseas Trading Co.
25th Sixth Avenue
New York, NY 10013

Dear Ms. Appleton:

I am writing concerning possible employment opportunities with your firm. In particular, I am looking for a senior MIS management position in a progressive international firm. By way of introduction, I have enclosed my resume.

As an MIS professional, I have had more than 15 years' experience developing large commercial systems. I possess a solid technical background in multi-language programming and systems design with extensive user interfacing.

Furthermore, I have direct experience in designing and establishing a systems programming organization for Deloitte & Touche to enhance the firm's internal business system. I would welcome the opportunity to assist in the enhancement of Overseas Trading Co.'s information systems.

I would appreciate an opportunity to discuss my background with you in greater detail. I look forward to hearing from you.

Sincerely,

Jason Dean

enclosure

Lloyd G. Prescott
5958 Ivy Drive
Newark, NJ 07430
(201) 555-5297

April 23, 20__

Mr. Sam Palmer
Personnel Director
Michigan Engineering
P.O. Box 250
Detroit, MI 48226

Dear Mr. Palmer:

The purpose of this letter is to ask for your firm's consideration for an available position as a Senior Industrial Engineer.

I have approximately 15 years' experience in industrial engineering and manufacturing process engineering with an emphasis on cost and manpower reductions, general floor troubleshooting, automation, equipment justification, and line balancing.

After you have reviewed my resume, I would appreciate the opportunity to discuss with you any industrial engineering openings. Thank you for your time.

I look forward to your response.

                                        Sincerely,

                                        Lloyd G. Prescott

February 3, 20__

Bechtel Laboratories
488 Industrial Parkway
San Francisco, CA 94105

ATTN: Rita Long, Director of Human Services

Dear Ms. Long:

Presently I am the project coordinator for the University of Arizona Physical Plant. My responsibilities include managing construction projects, estimating, surveying, and supervising jobs.

I am seeking a more demanding position with an international construction firm. My goal is to advance into a construction management position that focuses on achieving environmental compatibility in projects undertaken.

My resume is enclosed for your review. I look forward to hearing from you soon about career opportunities at Bechtel.

Sincerely,

Karen Adams
1685 Mountain Drive
Tucson, AZ 85720
(602) 555-8960

October 9, 20__

Columbus Engineering
200 East 16th Street
Columbus, OH 43216

Dear Mr. Johnson:

I am interested in interviewing with you for an entry-level civil engineering position.

Recently, I received my masters in civil engineering from Ohio State University. Furthermore, I passed the Engineer in Training Examination this past August.

While my work experience has been primarily in the structural analysis and design of bridges, roadways, and sewer systems, I am willing to consider positions in other related areas.

I have enclosed my resume for your review and consideration. I would like to meet with you personally to discuss my qualifications. I will follow up this letter with a telephone call in a few days.

Sincerely,

Mark F. Fulton
78 Prairie Road
Columbus, OH 43216
(614) 555-3981

enclosure

Fred Stevens
474 Drury Lane
Princeton, NJ 08544
(609) 555-3061

April 5, 20__

Mr. Richard Fraser
Western Telecom
3200 Valley Way
Englewood, CO 80111

Dear Mr. Fraser:

I am writing to obtain further information regarding employment opportunities with your corporation in the area of telecommunications research and development. Specifically, I am interested in pursuing a career in optical fiber networks, satellite communication, or antenna design.

I will be graduating from Princeton University in June with a master's degree in Electrical Engineering. My studies have been concentrated in telecommunications and fiber optics. In addition, my summer internship enabled me to conduct research on privacy and security issues in passive, cyber-to-the-home networks.

A copy of my resume is enclosed for your evaluation. If you need further information, I will be pleased to provide you with the necessary materials.

I look forward to discussing with you soon career opportunities at Western Telecom. I may be reached at the above number. Thank you for your consideration.

Sincerely,

Fred Stevens

June 15, 20__

Mr. Victor Lord, Senior Partner
Industrial Design Group, Inc.
488 West Lafayette Blvd.
Cuyahoga Falls, OH 44221

Dear Mr. Lord:

I am very interested in pursuing a designer/drafter position with Industrial Design Group, Inc. A copy of my resume is enclosed for your review.

Through my present employment with Commercial Design, I have refined my design and detail skills. My work with this firm entails design development of mechanical, electrical, and plumbing systems for major commercial projects. I have gained firsthand experience with developing the construction details of the engineering and architectural concepts as well as preparing the final bid documentation.

I look forward to hearing from you soon to further discuss my qualifications.

Sincerely,

Paul J. Richards
3300 Westwood Drive
Cuyahoga Falls, OH 44221
(216) 555-6929

EVAN LINDQUIST
6189 Beach Road, #9C
Jacksonville, FL 32209
904/555-1263

April 15, 20–

Jack Fang
General Aeronautics
100 Canaveral Road
Daytona Beach, FL 32014

Dear Mr. Fang:

This letter is in response to your advertisement in last Tuesday's *Miami Herald*. I am interested in the aerospace engineer position with your firm.

Currently I am working as a Strength Engineer for Aerospace International, where I am responsible for conducting detailed stress analysis for aircraft engine components. Previously, I have been a Value Engineer with Aircraft Technology Corporation working on aircraft air-conditioning systems and a Stress Engineer with Goddard Aerospace Systems working on various commercial aircraft. Based on my experience, I believe that I could make a valuable contribution to General Aeronautics.

I hope to further discuss my qualifications with you in an interview.

Yours truly,

Evan Lindquist

enclosure

May 28, 20__

Mr. Henry Hart
Carnegie Steel
Gary, IN 46408

Dear Mr. Hart:

I wish to apply for a position as an electrical engineer with Carnegie Steel. I possess bachelor's and master's degrees in electrical engineering. Also, I have worked for more than eighteen years as a steelmaking reliability supervisor and manager and as a steel operations maintenance engineer. This extensive experience coupled with my personal interest could be of value to your firm.

For your review, I am enclosing my resume. If you wish to set up an interview, please feel free to call me at (219) 555-6823.

Sincerely yours,

Scott Monroe
64 Fountain Lake Road
Gary, IN 46408

enclosure

March 30, 20__

Sandia Robotics
Human Resources
500 East 29th Street
Trenton, NJ 08391

SUBJECT: Opening in robotics lab for a programmer

To whom it may concern:

I am writing to you with the hope that you might have an opening soon in your robotics laboratory for a programmer. If you do not, I would appreciate your keeping my resume on file for upcoming opportunities.

My coursework for a bachelor's degree in mechanical engineering at Princeton University will be completed in 1994. Currently, I am performing independent robotics research on the use of Linear-Motor-robots in assembly as a result of my research fellowship.

People who know me well consider me to be dedicated, hardworking, and creative. I enjoy work and perform well under pressure. I believe that these characteristics fit with the type of professional you will be seeking.

Thank you for considering my qualifications. I look forward to hearing from you soon.

Sincerely,

Rachel Schwartzman

Patricia Butterfield
1480 Dean Road
Sacramento, CA 95819
916-444-2131

March 30, 20__

Ms. Consuelo Flores
EnviroTek
33 Sierra Road
Los Angeles, CA 90024

Dear Ms. Flores:

For almost ten years, I have pursued a satisfying career with the California Department of Environmental Management. At this point in my career, I would like to make a change to the private sector.

During my career, I have held positions as Environmental Project Manager, State Cleanup Section; Environmental Manager, Facilities Planning Section; and Environmental Scientist, Permits Section. My responsibilities have included managing the cleanup of hazardous waste sites, reviewing construction plans for water treatment facilities, and writing municipal permits. As a result, I have become more adept at maneuvering through the bureaucracy to make things happen quickly.

I would welcome the opportunity to speak with you about my background and the potential areas where my expertise could be best utilized by your firm. A resume is enclosed detailing my qualifications.

Sincerely,

Patricia Butterfield

September 27, 20__

Star Tribune
P.O. Box 744
Houston, TX 77022

SUBJECT: Opening for project geologist

Wednesday's ad for a project geologist prompted me to send this resume. I am interested in working in the Houston area. As you can see from my resume, I have more than four years of experience in managing environmental assessments as well as designing groundwater recovery and treatment systems.

I would like to meet with you to discuss my qualifications. I believe that I would be a productive member of your engineering staff.

Sincerely,

Bryan Pullman
43 Buffalo Bill Road
Omaha, NE 68129
(402) 555-5837

October 8, 20__

Ms. Millicent Jones
Hancock Construction
1501 Pennsylvania Ave., NW
Washington, D.C. 20006

Dear Ms. Jones:

I am writing to follow up our telephone conversation of yesterday morning regarding Hancock Construction's need for a project manager in Beijing. As we discussed, I am currently winding up a construction project for Johnson, Inc., and am seeking a project management position overseas. My work experience combined with my fluency in Mandarin Chinese uniquely qualifies me for this position.

If I meet the requirements, I would be available for employment at the start of the new year. Relocation to Beijing presents no problems for my family.

I have enclosed my resume as you requested. I am looking forward to hearing from you soon.

Sincerely,

John K. Lai
20 West Concord Street
Dover, NH 03820
(603) 555-1703

July 13, 20___

Southern Glassworks
350 Florida Street
New Orleans, LA 70112

Dear Personnel Manager:

This letter is in response to the advertisement in the *New Orleans Picayune* of last Sunday for an industrial engineer. Please accept my resume in consideration for this position.

With a master's degree in industrial engineering from Tulane University and seven years of work experience as an industrial engineer at Alexander Steel Company, I believe that I'm well-suited to your firm's needs.

Thank you for your time. I look forward to hearing from you soon regarding the position at Southern Glassworks.

Sincerely,

Edgar Peters
9 De Soot Drive
Baton Rouge, LA 70112
(504) 555-1388

enclosure

STEVEN R. KOHLHASE
473 HILL DRIVE WEST
KENOSHA, WI 53143
(414) 555-6320

June 17, 20__

Mr. Richard Serafini
Alistates Engineering
6341 Crestwood Dr., Suite 416
Naperville, IL 60665

Dear Mr. Serafini:

Given your company's excellent reputation in environmental engineering, hydrogeology, and solid waste, your firm must appreciate the need for polished, professional business writing for all the project reports you submit. My education in environmental science and work experience as a technical writer/editor with an engineering firm have given me the knowledge and solid writing skills that can benefit a firm like yours.

My enclosed resume will show you that I have a good background in public affairs. As you will also notice, my coursework in environmental chemistry, geology, and systems analysis have provided a strong foundation for my career focus on environmental writing. I am also very familiar with reading blueprints, reviewing cost estimates, change orders, and bidding procedures. I am hardworking and have consistently met publishing deadlines. In addition, I am a skilled electronics technician.

Because proven ability and skills are best evaluated in person, I would appreciate an interview with you. Thank you for your time and your consideration. I will be waiting to hear from you.

                    Respectfully yours,

                    Steven R. Kohlhase

GORDON EXTINE
2556 Broadlawn Street
Houston, TX 88674

May 27, 20__

Personnel Director
Dimetrics, Inc.
P.O. Box 788964
San Francisco, CA 94147

Dear Personnel Director:

I would like to be considered for an environmental position with Dimetrics, Inc. I have graduated from Notre Dame with a B.S. in Public Affairs. My majors were Environmental Science and Environmental Affairs.

I feel that my experience in the field of environmental science and as student assistant to the science department, along with my education, qualifies me for a position with your firm. I will continue to be successful as an environmental scientist because I enjoy challenges, hard work, and am concerned with doing the best that I can at all times.

I would like to request an interview to discuss how my placement with your firm will benefit both of us. Please phone me at (218) 555-8866. I look forward to hearing from you.

Yours truly,

Gordon Extine

FELICITAS A. FINNER

Telephone 415-555-9466
55778 Ventura Blvd.
Encino, CA 91319

July 2, 20__

Mr. David Mendoza
Director, Human Resources
Cimflex Corporation
P.O. Box 887
Topeka, KS 66608

Dear Mr. Mendoza:

Thank you for speaking to me by phone this afternoon. As you know, I am pursuing a career in robotics and would like to learn more about Cimflex Corporation.

In May of this year I graduated from Indiana University with a bachelor's degree in robotic engineering. My internship with Econtron Engineering offered me valuable exposure to the operations of a consulting firm involved in the robotics field.

The enclosed resume should assist you in evaluating my qualifications. If you need further information, please let me know.

I will look forward to meeting with you to discuss my qualifications for employment with your company. Thank you for your consideration.

Sincerely,

Felicitas A. Finner

Enclosure

Elizabeth A. Grossa
1346 E. 22nd St., #105
Carbondale, IL 62901
(217) 555-9142

May 26, 20__

Box 279
Indianapolis Star
457 Martindale Road
Indianapolis, IN 46227

APPLICATION FOR STRUCTURAL ENGINEERING POSITION

This letter is in response to the ad placed in this Sunday's edition of the *Indianapolis Star.*

I will earn my B.S. degree in Structural Engineering from Southern Illinois University this August. I have specialized in the fields of structural integrity and finite elements analysis. Your ad was of particular interest to me, as the job also dealt with some customer and sales interaction.

Enclosed is my resume detailing my work experience and educational background. I feel that my qualifications would be an asset to your organization.

I would welcome the opportunity to meet with you to discuss my experience and qualifications.

Sincerely,

Elizabeth A. Grossa

DAN LUI

17 Dinge Road
Terre Haute, IN 52211
317/555-1331
317/555-2339

August 21, 20___

Farrallon, Inc.
787 E. Fourier Drive
Emeryville, CA 96998

Attn: Robert Crain, Director of Human Services

Dear Mr. Crain:

After your visit to Rose-Hulman last March, we spoke about opportunities within your company for electrical engineers. You indicated that new positions would be opening this fall. I am writing to request an interview for one of those openings.

In May, I graduated from Rose-Hulman with a B.S. in Electrical Engineering. I was one of fifteen out of two hundred who graduated with honors. My coursework included microwave circuit design, electromagnetic waves, digital integrated circuits, and systems and signals.

As I look forward to my career in this field, I know that I would be able to make good use of my education working for Farrallon.

I have enclosed a copy of my resume and will call next week to discuss setting up an interview.

Sincerely,

Dan Lui

2556 Forest St.
Houston, TX 77063

January 27, 20__

Personnel Director
Valley Hospital
P.O. Box 228964
Birmingham, AL 35222-8964

Dear Personnel Director:

Please consider my application for a position as a dietitian. I graduated from Purdue University with a Bachelor of Science degree in dietetics. I have been a registered dietitian for the past three years.

I feel that my experience as a dietitian in a nursing home and clinic, along with my education, qualifies me for a position with Valley Hospital. I will continue to be successful as a dietitian because I enjoy the challenge of helping people regain their health through proper diet. Furthermore, I work hard and am concerned with doing my best at all times.

I would like to request an interview to discuss how my placement with your hospital would be of mutual benefit. Please phone me any time at (713) 555-8866. I look forward to hearing from you.

Yours truly,

Edda Fisher

Enclosure